Jigar Chaudhary
Sushilkumar Roy
Ajay S. Patel

Buffalo Milk Sandesh

Jigar Chaudhary
Sushilkumar Roy
Ajay S. Patel

Buffalo Milk Sandesh

LAP LAMBERT Academic Publishing

Impressum / Imprint
Bibliografische Information der Deutschen Nationalbibliothek: Die Deutsche Nationalbibliothek verzeichnet diese Publikation in der Deutschen Nationalbibliografie; detaillierte bibliografische Daten sind im Internet über http://dnb.d-nb.de abrufbar.

Bibliographic information published by the Deutsche Nationalbibliothek: The Deutsche Nationalbibliothek lists this publication in the Deutsche Nationalbibliografie; detailed bibliographic data are available in the Internet at http://dnb.d-nb.de.

Coverbild / Cover image: www.ingimage.com

Verlag / Publisher:
LAP LAMBERT Academic Publishing
ist ein Imprint der / is a trademark of
OmniScriptum GmbH & Co. KG
Heinrich-Böcking-Str. 6-8, 66121 Saarbrücken, Deutschland / Germany
Email: info@lap-publishing.com

Herstellung: siehe letzte Seite /
Printed at: see last page
ISBN: 978-3-659-58548-7

Brief Information

Sandesh occupies a prominent place among indigenous milk products and carries lot of market potential. Sandesh is the most popular chhana based sweet delicacy of the eastern part of India particularly in West Bengal and Bangladesh. It is well known that buffalo milk possess certain problems during its conversion into particular milk product. Buffalo milk chhana lead to hard & coarse texture Sandesh, which are considered as defect in Sandesh. Certain alteration is necessary in the processing technique for conversion of buffalo milk to obtain good quality chhana thereby good quality Sandesh. Some of the treatments which can be adopted are adjustment in fat percent of milk, adjustment in salt balance of buffalo milk, homogenization of milk, change in the coagulation temperature, etc. In India, buffalo milk account for over 55 percent of the country's total milk production. So, the main source of marketable surplus is buffalo milk. In recent years, there has been growing interest in the manufacture of Sandesh and other chhana based product from cow and buffalo milk.

The present study has been carried out to study the effects of processing parameters (fat percent of milk, adjustment in salt balance of milk, homogenization of milk, coagulation temperatures ect.) on Sandesh made from buffalo milk. The final product made from buffalo milk using with various suitable processing parameters were also evaluated for sensory, chemical, microbial and rheological attributes.

A minimum fat content of 5% in buffalo milk was essential to obtain Sandesh with satisfactory body and texture. Addition of sodium citrate to different level (0.2% and 0.3%) prior to its heating and coagulation resulted in major enhancement in body and texture of buffalo milk Sandesh, though there was a problem of fat leakage. The product obtained at 0.3% was relatively more acceptable than at 0.2%.

Homogenization of buffalo milk in two stages at 1500 and 500 psi resulted in better body and texture without fat leakage and also it was close to control, whereas product obtained from buffalo milk homogenized at 1000 and 500 psi resulted in poor body, texture and overall acceptability.

Out of the two final heating temperatures, 90°c resulted in products with somewhat better overall acceptability than products at 95 °c. The body and texture and yield of Sandesh samples made from buffalo milk coagulated at 70 °c were better than coagulated at 80 °c. There was a little marked affect on the flavour and colour and appearance of the product due to temperature of coagulation. The mixture was cooked at 70 °c for 20 minutes resulted in soft grade Sandesh, whereas the mixture was cooked at 80 °c for 20 minutes gave comparatively hard, brittle and chewy product.

Sandesh made from control recorded more total solid and titratable acidity as compare to that of buffalo milk Sandesh. The fat and carbohydrate content decreased but the protein and ash contents increase slightly in the product from buffalo milk Sandesh.

There were significant reduction in the hardness, cohesiveness, springiness, gumminess and chewiness of the final product as compare to that of cow milk Sandesh.

Finally it was concluded that Sandesh made from buffalo milk using various suitable processing parameters gave slightly better body and texture than that of cow milk Sandesh, whereas the colour and appearance was comparatively inferior.

CONTENTS

SrNo.	Chapter	Page No,
I.	INTRODUCTION	7-9
II.	REVIEW OF LITERATURE	10-25
III.	MATERIAL AND METHODS	26-36
IV.	RESULTS	37-61
V.	DISCUSSION	62-66
VI.	SUMMARY AND CONCLUSION	67-69
	BIBLIOGRAPHY	70-76

LIST OF TABLES

Table No.	Title
2.1	Chemical Composition of Chhana Obtained from Various Sources
2.2	Average chemical composition of Sandesh
3.1	Classical 9 point hedonic scale method
4.1	Effects of fat level
4.2	Effects of sodium citrate
4.3	Effects of homogenization
4.4	Effects of final heating temperature
4.5	Effects of coagulation temperature
4.6	Effects of cooking temperature
4.7	Sensory evaluation of final product
4.8	Chemical composition of control and Final
4.9	Microbial counts of Control and Final product.
4.10	Rheological values of control and Final product.

LIST OF FIGURES

Table No.	Title
3.1	A typical profile analysis curve of Sandesh
3.2	Photograph of food texture analyzer of Lloyd Instruments LRX plus material testing machine, England
4.1	Photograph showing effects of fat level on buffalo milk Sandesh
4.2	Photograph showing effects of sodium citrate on buffalo milk Sandesh
4.3	Photograph showing effects of homogenization on buffalo milk Sandesh
4.4	Photograph showing effects of final heating temperature on buffalo milk Sandesh
4.5	Photograph showing effects of coagulation temperature on buffalo milk Sandesh
4.6	Photograph showing effects of cooking temperature on buffalo milk Sandesh
4.7	Photograph showing all finally selected parameters on buffalo milk sandesh
4.8	A typical texture profile analysis curve of control
4.8	A typical texture profile analysis curve of final product

LIST OF ABBREVIATIONS

%	Percentage
BIS	Bureau of Indian Standards
^{0}C	Degree centigrade
et al.	et alibi
g	Gram
gf	Gram force
hr	Hour
IU	International unit
kg	Kilogram
mg	Milligram
ml	Milliliters
mm	Millimeter
N	Normal (normality)
N	Newton
pH	The negative logarithm base ten of the hydrogen-ion concentration
PSI	Pounds per square inch
PFA	Prevention of Food Adulteration Act
rpm	Revolutions Per Minute
RH	Relative Humidity
SNF	Solid –not –fat
ug	Microgram
v/v	Volume by Volume
w/w	Weight by Weight

CHAPTER I

INTRODUCTION

The Milk and milk products are an integral part of human nutrition. The country has predominant agrarian population, therefore milk and milk product have a significant place in the people's diet in general and sick people in particular. Milk is said to be the most complete food because of its biological value as it contains a variety of nutrients and therefore it is rightly said "Milk is nature's most nearly perfect food."

Historically, surplus milk in the rural areas where it is produced has been converted into a variety of traditional products primarily as a means of preservation. The increased availability of milk during the flush season coupled with lack of facilities to keep liquid milk fresh during transit from rural production areas to urban market makes conversion of milk into traditional products is a compulsion. About 50% of milk produced in India is converted by the traditional sector (halwais) into a variety of traditional milk products (Dairy India, 2007). Traditional dairy products refer to those milk products which originated in undivided India (this includes India, Pakistan and Bangladesh) (De, 1980).The preservation of milk by beneficial micro-organisms, concentration/preservation have resulted in creation of a large number of milk delicacies. The traditional milk products have been classified into three main classes:

(a) Heat desiccated milk products like Khoa, Peda, Barfi etc.

(b) Acid precipitated milk products like Paneer, Chhana etc.

(c) Cultured milk products like Dahi, Shrikhand etc (Aneja, 1997).

Under acid precipitated milk product, chhana and paneer are the main products. Paneer is used in the preparation of culinary dishes whereas chhana is used for sweetmeat preparation. Chhana, a well known traditional indigenous milk product is obtained by acid coagulation of hot milk and is used extensively as a base material for large variety of Indian delicacies namely Sandesh, rasogolla, cham-cham, rasmalai and many other such product. Sandesh is a very popular and one of the oldest chhana based sweetmeat (Aneja, 1987). It has been estimated that the annual production of Sandesh in West Bengal alone is 30,000 tons (Bandyopadhyay and Khamrui 2007). The popularity of this sweetmeat is gradually spreading to the other part of the country (Rajorhia and sen, 1987). Sandesh has yellowish white to pale yellow colour, a sweet, pleasant, slightly cooked, caramelized flavour with soft and cohesive body, smooth texture with small size grains (Sen and Rajorhia, 1985; Sen and Rajorhia, 1991; Aneja, 1997). It is a rich source of milk protein, fat, carbohydrate and the vitamins like *A* and *D* (Sahu and Jha, 2008). Three varieties of Sandesh recognized which are soft grade (norampac/ batapak/babupak), hard grade (karapak) and kachhagolla (Sen and Rajorhia, 1985).

The yield and quality of chhana for the preparation of Sandesh depend on several factor viz. type of milk, heat treatment prior to acidification, coagulation temperature, acidity of milk, concentration of coagulant and residual time of coagulation before separation (Jank man & Das, 1993 and choudhary *et al.*, 1998). Cow milk chhana is usually preferred for preparation of Sandesh as it gives a product with soft body, smooth texture with uniform grains (De and ray, 1954). Buffalo milk on the other hand tends to produce a hard body and coarse texture in Sandesh due to its higher concentrations of proteins and minerals (Sanyal *et al.*, 2011).

In India, buffalo milk account for over 55 percent of the country's total milk production (Dairy India, 2007). So, the main source of marketable surplus is buffalo milk. Keeping in view, the increased production of buffalo milk and surplus availability in the organized dairy/ unorganized sector, its utilization in the manufacture of various products is gaining momentum. It is well known that buffalo milk possess certain problems during its conversion into particular milk product. Buffalo milk chhana lead to hard & coarse texture Sandesh, which are considered as defect in Sandesh (Sen and Rajorhia, 1985; Sen and Rajorhia, 1990; Sen and Rajorhia, 1991 and Das, 2000). In addition, lower hydration capacity of buffalo milk casein and its higher calcium content further impart the quality of buffalo milk chhana.

Earlier, few attempts were made to improve the quality of chhana from buffalo milk. (Jagtiani *et al.*, 1960 and Kundu and De, 1972). Consequently, certain alteration is necessary in the processing technique for conversion of buffalo milk to obtain good quality chhana thereby good quality Sandesh. Some of the treatments which can be adopted are adjustment in fat percent of milk, adjustment in salt balance of buffalo milk prior to its heating and coagulation, homogenization of milk, change in the coagulation temperature, etc.

Therefore, it is worthwhile to see the possibility of preparing Sandesh from the modified buffalo milk chhana with undesirable change. Looking to the facts & figure mentioned above, the present study is undertaken.

1.1 Objective:

To obtain the requisite quality chhana Vis - a- Vis Sandesh, the following objectives were undertaken:

1 To develop chhana vis - a- vis Sandesh from buffalo milk.

2 To evolve processing practice under cottage / industrial situation.

3 To carry out the proximate analysis of developed product.

CHAPTER II

REVIEW OF LITERATURE

Sandesh occupies a prominent place among indigenous milk products and carries lot of market potential. Sandesh is the most popular chhana based sweet delicacy of the eastern part of India particularly in West Bengal (and some parts of Bihar, Assam, Tripura, Orissa) and Bangladesh. Sandesh is prepared by kneading chhana into uniform dough, mixed with sugar (30-40% of chhana) and cooked over low flame with constant scraping until the mixture gets the desired consistency and flavor. Sandesh is popular due to its palatability and aroma. Three distinct varieties of Sandesh are popular: soft grade (norampac), hard grade (karapak) and kachhagolla. The most common variety is the soft grade Sandesh. It has soft body and smooth texture with fine grains uniformly distributed.

In recent years, there has been growing interest in the manufacture of Sandesh and other chhana based product from cow and buffalo milk. The relevant information has been reviewed as follows:

2.1: Importance of Chhana

Chhana is defined as a milk product obtain by precipitating a part of milk solid by boiling whole milk of cow and or buffalo or a combination thereof by addition of lactic acid, citric acid or other suitable coagulating agent and subsequent drainage of whey. And it should not contain more than 70% moisture and milk fat should not be less than 50% of the dry matter (PFA, 1976). Chhana is used as a base material in the preparation of variety of sweets like Sandesh, rasgolla, pantooa and chhana murki etc.

Pattern of milk consumption in India indicate that about 6% of milk is coagulated for production of chhana (Sahu and Das, 2007). It has been reported that in India, the market volume of chhana-based sweets is about 1 million tones with a value of Rs 7,00,000 crores (Sahu, 2007). It has been also estimated that about 80% of chhana produced is converted into Sandesh (Aneja, 1997). Chhana is a rich source of fat and protein. It also contains fat soluble vitamins A and D with high protein and low sugar content, chhana is highly recommended for diabetic patients (De, 1980). Since buffalo milk constitute more than 55 per cent of milk production in India, it is worthwhile to see the possibility of preparing chhana based sweets from buffalo milk.

2.2: Prepation of Chhana

2.2.1: Selection of Milk

Cow milk chhana is usually preferred for preparation of Sandesh as it gives product with soft body, smooth texture with uniform grains and mildly acidic flavor (De and ray, 1954). It is well known that buffalo milk possess certain problems during its conversion into particular milk product. Buffalo milk chhana lead to hard & coarse texture Sandesh, which are considered as defect in Sandesh (Sen and Rajorhia, 1985; Sen and Rajorhia, 1990; Sen and Rajorhia, 1991 and Das, 2000).

Buffalo milk chhana is not preferred to cow milk chhana because of higher concentration of casein additional in micelles state with bigger size of the micelles, harder fat due to large portion of high melting triglycerides with bigger size of fat globules and higher content of calcium more so in the colloidal state may be responsible for harder and less cohesive chhana (Sindhu *et al.*, 2000). Therefore, certain modification is necessary in the processing technique for conversion of buffalo milk so as to obtain good quality chhana. Some of the treatments which can be adopted are alteration in salt balance of buffalo milk prior to its heating and coagulation, homogenization of milk, change in the coagulation temperature, reducing

SNF content by dilution etc. Earlier, few attempts were made to improve the quality of chhana from buffalo milk. Some reports indicate modification of buffalo milk for chhana production (Jagtiani *et al*., 1960 and Kundu and De, 1972).

A minimum fat content of 4% in cow milk and 5% in buffalo milk was essential to obtain chhana with satisfactory body and texture (De, 1980). Crossbred cow's milk having 4% fat resulting in best quality chhana for rosagolla making (Rao, 1971). Lower than 4% fat lead to hard body and coarse texture in chhana while higher fat levels result in the greasy surface (Rajorhia and Sen, 1988). When calcium was present above a particular limit in milk, it rendered chhana hard. Buffalo milk contains about 0.21% calcium whereas cow milk contains 0.12% calcium. Thus by adding 0.2-0.3% sodium citrate as a softening agent in buffalo milk and storing such milk for some time before precipitation helped in producing soft chhana (Jagtiani *et al.,* 1960). Addition of sodium dihydrogen phosphate, disodium dihydrogen phosphate, sodium citrate and their combination in veering quantities ranging from 1 to 2 % in buffalo milk prior to coagulation help to produce soft product (De and Ray, 1954 and Date *et al.,* 1958).

Dilution of buffalo milk with 25% water prior to coagulation improved the softness of chhana (Aneja *et al.,* 1982).The treatment of buffalo milk with 0.05% sodium citrate prior to boiling, dilution with 25% water and coagulation with 1% citric acid solution improved the quality of chhana to a considerable extent (Iyer, 1978).Curd tension of milk influences the body and texture of chhana. Used of 0.25 to 0.30% sodium citrate to replace calcium, which in turn gave reduce curd tension to improve the quality of chhana and rasogolla made from buffalo milk (Jagtiani *et al.,* 1960). Homogenization of buffalo milk at 2500 psi would improve the softness of chhana. Also homogenization of milk and delay staining increase both yield and present recovery of milk solid in buffalo milk chhana (Kundu and De, 1972). A homogenization pressure of 1900 psi should be adequate to improve the body and texture of buffalo milk chhana (Kanawjia, 1975 and Gajendra, 1976). On the contrary, the same treatment did not give any beneficial effect of homogenization of buffalo milk for chhana making (Iyer, 1978; Soni *et al.*, 1980 and Ahmed *et al.,* 1981). Adulteration of milk with starch tends to produce gelatinous mass, and presence of colostrums in milk lead to a pasty texture in chhana, both of which are unsuitable for sweet making (Ray and De, 1953).

2.2.2: Final Heating Temperature of Chhana

Generally for chhana preparation, milk is heated nearer to boiling point. For the preparation of good quality Sandesh, milk is heated to 90-95^{0}C, followed by cooling to 70^{0}C (Sahu and Jha, 2008). For good quality chhana preparation, milk was heated to 90^{0}c and then coagulated at 70^{0}c (Das, 2000).Continuous heat-acid coagulation unit ware used by Sahu and Das, (2009), according to them final heating temperature was 95±2°C.

2.2.3: Coagulation of Milk

The quality of chhana is greatly influenced by type of coagulant, temperature of coagulation residence time after precipitation of the milk (Singh and Ray, 1977a and Rajorhia and Sen, 1988). Coagulation of milk in chhana making is due to combined effect of chemical and physical changes in the casein micelles brought about by the action of acid aided by relatively higher temperature of coagulation (Davies, 1948). Acid affects the stability of casein directly by disturbing the charges carried by particles and indirectly by releasing the calcium ion from colloidal calcium-caseinate-phosphate complex (Sahu and Das, 2009).

2.2.3.1: Type of Coagulant

Generally organic acid like citric, lactic or their salt (calcium lactate), lemon juice and sour whey are employed as coagulants for chhana preparation. Lactic acid tends to produce chhana with granular texture (fit for rosogolla making), while citric acid results in pasty texture (fit for Sandesh making) (Ray and De, 1953 and De and Ray, 1954). Sour whey with about 0.9% acidity may be used for chhana preparation which was suitable for rosogolla making (Mahanta, 1964; Srinivasan and Anantakrishna, 1964; Gera, 1978 and Aneja *et al.,* 1982), whereas good quality chhana was obtained from sour whey of 1.6% acidity (Singh and ray, 1977).

The use of calcium lactate as a coagulant for chhana making at home is very common in West Bengal (Chakravarti, 1982). Calcium lactate produces chhana with bright white colour, soft body, smooth texture and pleasant flavor. This chhana can be used for Sandesh preparation (Sen and De, 1984). For cow milk, chhana derived from 2% calcium lactate solution was unsuitable for Sandesh preparation, whereas chhana derived from 4 and 6% calcium lactate solution was suitable for Sandesh preparation (Sen, 1985).

2.2.3.2: Strength of Coagulant Solution

Low acid strength(0.5%) result in very soft body and smooth texture suitable for rosogolla but unsuitable for Sandesh making (De and Ray,1954 and Soni *et al.*,1980), while high acid strength results in hard body and less smooth texture, suitable for Sandesh making but not for rosogolla. The optimum strength of coagulant solution should be between 1 and 2% citric or lactic acid to produce good quality chhana suitable for making both kinds of sweetmeat (De and Ray, 1954 and Iyer, 1978).

Some workers have recommended that good quality chhana from buffalo milk can be prepared from 1% citric acid solution (Kundu and De, 1972 and

Ahmed *et al.*, 1981). A satisfactory quality of chhana from buffalo milk was also prepared with 0.5% lactic acid coagulant (Soni *et al.,* 1980). Chhana produce in continuous heat-acid coagulation unit where citric acid strength was 1.62% (Sahu and Das, 2009).

2.2.3.3: Amount of coagulant

Usually, 2-2.5g of citric or lactic acid per kg of fresh milk is needed for coagulation. About 2.5-3.4 g of citric acid or 3.0-3.9 g of lactic acid was necessary to achieve complete coagulation (Iyer, 1978). The exact quantity of coagulant is dependent on type of milk. About 1.5 g of citric acid required per kg of buffalo milk (Gajendra, 1976) whereas only 1.25 g of lactic acid per lit. of buffalo milk is required to produce good quality chhana for rosogolla (Soni *et al.*, 1980). The amount of calcium lactate needed for complete coagulation ranged from 6 to 12 g per kg of milk depending on the coagulation temperature (Sen, 1986).

2.2.3.4: Coagulation Temperature

Chhana of satisfactory quality from cow milk can be made when milk is coagulated at 82^0c and the coagulation process completed in 0.5-1.0 min (De and Ray, 1954), whereas the best quality chhana from cow milk was obtained at a coagulation temperature of 70°c (Iyer, 1978). Coagulation of cow milk at 80^0c was found to be optimum for rosogolla making (De and Ray, 1954; Bhattacharya and Desh raj, 1980 and Aneja *et al.*1982). The optimum coagulation temperature for chhana making from buffalo milk is 70°c (Kundu and De, 1972; Gajendra, 1976; Iyer, 1978; Soni *et al.*, 1980 and Ahmed *et al.*, 1981).

The amounts of coagulant require for completing the coagulation of milk is increased with the lowering of coagulation temperature (Singh and ray, 1977; Soni *et al.,* 1980 and Sen., 1986). As coagulation temperature decrease, the moisture retention in chhana increase leading to its softer body and smooth texture (De and Ray, 1954; Soni *et al.,* 1980 and Sen., 1986).

2.2.4: pH of Milk

The optimum pH for chhana making from cow and buffalo milk is 5.4 and 5.7 respectively. The pH of coagulation principally regulates the moisture content and the body and texture which are the best obtained at the above pH (De and Ray, 1954; Singh and ray, 1977 and Soni *et al.,* 1980). Most suitable pH for coagulating all type of fresh milk is 5.1 (Iyer, 1978). An optimum pH of 5.85 for cow milk when calcium lactate was used as coagulant reported by Sen and Rajorhia, (1986).

2.2.5: Speed of Stirring During Coagulation

Higher speed of stirring during coagulation reduces the moisture content in chhana and increases its hardness. Slow stirring (40-50 rpm) is preferred to avoid foam formation (Ray and De, 1953 and De and Ray, 1954).

2.2.6: Method of Straining

The acidified milk is incubated for some time before separating the whey from the coagulum. In the traditional method of chhana manufacture, the milk acid mixture is strained through a muslin cloth and the product is hanged for the gravity drainage of whey for 20 minutes and the resultant chhana is used for the preparation of Sandesh (Jank man & Das, 1993). Delayed straining produce comparatively soft and smooth texture chhana than immediate straining. Delayed straining gives a higher proportion of moisture, yield, recovery of milk solids and lower hardness value in chhana than immediate straining (Rajorhia and Sen, 1988).

Immediate straining of whey was advocated for making rosogolla from cow milk (Goel, 1970; Singh and Ray, 1977; De, 1980; Bhattacharya and Des raj, 1980; Wane, 1992 and Rahate, 1993). Many workers have suggested delayed straining process especially for buffalo milk chhana production (Iyer, 1978; Son *et al.*, 1980 and Ahmed *et al.,* 1981).

2.2.7: Yield of Chhana

In general the outturn of chhana increases with lowering the strength of coagulant solution, coagulating temperature and spread of stirring during coagulation (De and Ray, 1954; Sen 1985; Sen, 1986 and Ahmed *et al.,* 1981). The yield of the chhana from buffalo milk is higher than cow milk (Ray and De, 1953). The average yield of 16.4% from cow milk and 22.5% from buffalo milk was reported by Ray and De, (1953), whereas the yield of 20% from cow milk and 25.8% from buffalo milk was reported by (Iyer, 1978).

The highest yield was observed at 70^0c coagulation temperature accounted by Kundu and De, (1972). According to them this peak yield was due to the combine effect of higher moisture and milk solid retention in chhana, but Iyer, (1978), described that the maximum yield at 70^0c is due to retention of higher amount moisture and not because of greater recovery of milk solid in chhana. Further, lowering of coagulating temperature produce an adverse effect on the yield of chhana (Kundu and De, 1972 and Iyer, 1978). Chhana made from calcium lactate showed always higher yield than citric acid. This was mainly due to higher moisture retention capacity in calcium lactate chhana (Sen and De, 1984 and Sen, 1986).

Yield of chhana obtained from the continuous heat-acid coagulation unit was 0.2027 kg/ kg milk (Sahu and Das, 2009). Homogenization of milk and delayed straining increase both yield and percent recovery of milk solid in buffalo milk chhana (Kundu and De, 1972).

2.2.8: Chemical Composition and Nutritive Value of Chhana

According to both PFA (2006) and BIS (1969) chhana should not contain more than 70 per cent moisture and milk fat should not be less than 50 per cent of the dry matter. Other scientist revealed that the type of coagulants did not have any appreciable effect on the composition of chhana (Singh and ray, 1977 and Kawal, 1979), while significant difference was found in moisture content between calcium lactate and citric acid chhana (Sen,1986). Chhana retains about 90% of fat and protein, 50% ash and 10% lactose of the original milk. The energy value of cow milk chhana ranges from 2866 to 3748 calories per kg. Chhana also retains appreciable proportion of fat soluble vitamin like A and D (Ray and De, 1953).

The average calcium, phosphorus, vitamin A, B1, B2 and Content per 100g of chhana sample were 208 mg, 138 mg, 366 IU, 73 ug, 15 ug and 2.8 mg respectively (Mini *et al.*, 1955). Because of possible losses in whey, chhana was a poor source of vitamins B, lactose, ascorbic acid and vitamin A contents. The loss of ascorbic acid during chhana making was about 57% as compared with the figure of boiled milk (Rajorhia and Sen, 1988).

Table 2.1: Average Chemical Composition of Chhana Obtained from Various Sources

Source	Constitutes (%)					Reference
	Moisture	Fat	Protein	Lactose	Ash	
Cow milk	53.4	24.7	17.6	2.2	2.1	Anantakrishnan and Srinivasan (1964)
Buffalo milk	51.6	29.6	14.5	2.4	1.9	Anantakrishnan and Srinivasan (1964)

Laboratory	58.8	22.9	17.3	-	-	Singh and Ray (1977)
Goat milk	55.3	23.5	17.2	2.2	1.6	Jagtap and Srinivasan (1973)
Cow milk	53.1	24.8	17.8	2.2	2.1	Kumar and Srinivasan (1982)
Buffalo milk	51.7	29.7	14.4	2.3	1.9	Kumar and Srinivasan (1982)
Market	62.3	16.1	17.1	2.2	2.2	Kumar and Srinivasan (1982)
Cow milk	55.7	23.5	16.4	2.2	2.0	Sen (1986)
Buffalo milk	54.9	24.3	17.3	2.1	2.2	Singh (1994)

2.2.9: Microbial Quality

The microbiological studies were carried out to assess the standard of cleanliness during production, packaging, transportation, storage and for ascertaining the shelf life of chhana. Market sample were heavily contaminated with a variety of organism. The average initial number of viable organisms per g of chhana was about 16000 which increased to 110 million after 48 hr (Anonymous, 1955-56). Few studies on type of organisms present in chhana samples revealed the presence of micrococcus, spore and nonspore forming rods contributing 45, 34 and 21% of total population respectively (Rajorhia and Sen, 1988).The most common type of moulds contaminated chhana samples belonged to the genera Penicillium, Aspergillus, Mucor, Rhizopus and Fusarium (Anonymous, 1955-56).

2.2.10: Shelf Life of Chhana

Chhana is a high moisture product. It is common knowledge that it does not keep longer than a day at room temperature. Average shelf life of chhana is about 3 and 12 days at 24^0c and 7^0c, respectively (De, 1980). Freshly made chhana wrapped in vegetable parchment paper could be preserved in good condition for 3-4 days at 21c ^{0}-27^0c and for about 10 days in a refrigerator (Srinivasan and Anantakrishna, 1964). The shelf life of laboratory sample of chhana ranged from 24 to 48 hour while for market chhana it was only 6 to 8 hour in summer and 16 to 20 hour in winter (Anonymous, 1955-56).

Although the addition of 0.5% sodium benzoate or 0.2% sodium propionate was effective to check the growth of microorganisms, but this preservative impaired disagreeable odours to chhana. The incorporation of propionate results in the improvement of keeping quality of chhana and reduces the disagreeable odours considerably (Rajorhia and Sen, 1988).

2.3: Preparation of Sandesh

Sandesh is a sweet product mostly produced in unorganized small-scale sectors wherein variation in quality between batches, days of production and shops are noticed (Patil, 2005). Three varieties of Sandesh recognized are soft grade (norampac/ batapak/babupak, hard grade (karapak) and kachhagolla (Sen and Rajorhia, 1985). Sandesh is prepared by kneading chhana into uniform dough, mixed with sugar (30-40% of chhana) and cooked over low flame with constant scraping until the mixture gets the desired consistency and flavor (Aneja *et al*. 2002).

The chhana is mix with sugar (30-40% of chhana) and the mixture is cooked at 70-80^{0}c for 10-20 minute. The product is then transferred into moulds or the product is set in tray (Sen and Rajorhia, 1985). For Sandesh making, citric acid chhana was better than lactic acid chhana. But in the trade practice, use of sour/aged chhana whey is most common than citric acid or lactic acid due to easy available and inexpensive nature of whey (Singh and ray, 1977a).

2.3.1: Mixing and Kneading of Chhana with Sugar

It is very important to mix the sugar with the chhana before cooking of the mixture. Cane sugar at the rate of 30 percent of the weight of chhana is added for the preparation of Sandesh was reported (Das, 2000). Mixing and kneading of sugar with chhana is done manually in traditional method of Sandesh manufacture (Sen and Rajorhia, 1985; Tarafdar *et al.*, 1988; Sen and Rajorhia, 1991 and Das, 2000), this method is unhygienic, inefficient and also time consuming. The final quality of the product also depends to a great extent on proper mixing and kneading of chhana sugar mixture. The use of disc grinding for the mechanical kneading of the chhana for rosogolla making was also studied (Tarafdar *et al.*, 1988).

2.3.2: Cooking of Chhana-Sugar Mixture

The main process of Sandesh preparation is the cooking of chhana sugar mixture to reduce its moisture and to obtain the characteristic flavor and texture. In the traditional method of manufacture, the chhana sugar mixture is cooked in a karahi kept directly over fire at 75 to 85^{0}c for 15 to 25 minutes depending upon type of Sandesh. The cooking at higher (>85) temperature gave hard, brittle and chewy product with little cohesiveness (Sen and Rajorhia, 1985 and Das, 2000).

Double jacketed steam kettle was used for the cooking of the chhana-sugar mixture (Sen and Rajorhia, 1990 and 1991), chhana was added in two lots during the cooking process. The whole part of sugar added with one lot of chhana and the temperature was then gradually rise to70^0c in 15 minutes by continuous stirring and scraping with the help of a wooden ladle. At this juncture, the mixture developed its initial pat stage and the remaining lot of chhana was added. Due to this addition, the temperatures of mixture suddenly reduced to about 46^0c and it was then slowly heated to about 60^0c in 10 minutes with continuous stirring and scraping.

2.3.3: Cooling and Shaping of Sandesh

When the mixture attains the desired consistency during cooking, the mixture is transferred to a shallow tray and allowed to cool and set. Hard grade Sandesh is usually molded at around 50^0c while the soft grade Sandesh is molded at room temperature (Sen and Rajorhia, 1985). The product is cooled slowly to 37^0c in about 10 minutes, shaped using different moulds and cut into different sizes with a sharp knife (Sen and Rajorhia, 1990 and 1991).

2.3.4: Quality of Sandesh

The quality of Sandesh is largely depending upon the ingredient used and method of preparation of the product. The chemical composition, sensory quality and microbial quality of market product vary greatly as these aspects depend on quality of chhana, way of preparation, hygienic condition, level of sugar addition and other additives used in the product. Acceptability of Sandesh during storage reduced mainly due to flavor deterioration (Sen and Rajorhia, 1990). The samples during storage showed a progressive increase in total acidity, free fatty acid (FFA), Peroxide value and free fat contents.

Sandesh has a limited shelf life of two or three days due to its high moisture content, unhygienic condition during manufacturing and handling.

Shelf life of Sandesh varies from one to six days depending upon the quality of chhana, type of Sandesh, method of handling and season (Sen and Rajorhia, 1985).

2.3.5: Chemical Composition

The chemical composition of tow most common varieties (hard grade and soft grade) of Sandesh sold in Kolkata market was determined by Sarkar, (1975). According to him the chemical composition of soft and hard grade of Sandesh showed wide variation with respect to moisture, protein and sucrose contents. Some minor differences in the chemical composition of Sandesh prepared from three different coagulant viz., citric acid, lactic acid and sour whey and compare this with those of market samples. The market samples of Sandesh contain high sugar, low fat and protein contents as compared with laboratory made Sandesh (Singh and Ray, 1977a).

The deteriorative changes occurred during storage of Sandesh had studied by Sen and Rajorhia, (1990) and according to them the common chemical deteriorative factors during storage were oxidation, lipolysis and acid development. Addition of sorbic acid in Sandesh to enhance the keeping quality of product but sorbic acid imparted its own typical flavor defect in product (Sen and Rajorhia, 1997).

Table 2.2: Average Chemical Composition of Sandesh

Constituent %	**Sandesh**	
	Cow milk	Buffalo milk
Moisture	25.50	27.14
Fat	19.89	18.82
Protein	18.48	18.71
Sugar	34.47	33.83
Ash	1.66	1.90

Ref. Verma (1997)

2.3.6: Microbial Quality

It has been estimated that about 20% of the total food supply in the world is wasted owing to microbial spoilage (Tilbury, 1980). Sandesh by virtue of its lower moisture and higher sugar content is expected to be less susceptible to microbial spoilage (Sahu and Jha, 2008). Low microbial count in hard grade Sandesh is expected due to lower moisture and high sugar contain as compared with other two varieties. Coliform and spore counts in hard grade Sandesh were almost negligible. On the contrary, higher proportion of moisture and lower amount of sugar tremendously accelerate the growth of microorganism in soft and kachhagolla varieties of Sandesh (Sen and Rajorhia, 1986). Few attempts were made to examine the product for microbiological quality with a view to assessing the standard of cleanliness during production, transportation, storage and handling. Microbial analysis of market sample of Sandesh indicated the presence of bacteria and coli forms.

Examination of Sandesh sample were carried out for total plate/viable count (TVC), coli form count, staphylococcal count, yeast and mould count, The counts per gram samples ranged from 0-1 x 10^5, 0-55 x 10^1, 0-49 x 10^2 and 0-35 x 10^2 respectively (Singh and Mukhopadaya, 1975). Suitability of some packaging materials for Sandesh was observed that total viable count (TVC) and staphylococcus counts had major increase over spores and yeast and mould counts at different storage condition in the experiment. The use of sorbic acid ceases the microbial deterioration (Sen and Rajorhia 1997).

2.3.7: Sensory Quality

The sensory properties play a significant role in the acceptability of the product. The 9 - point hedonic scale, given by Amerine *et al.*, (1965), is generally used to assess the overall liking and disliking of food products. The sensory characteristics included in the hedonic scale are flavor, body and texture, colour and appearance and overall acceptability. Sensory evaluation revealed that the scores of the Sandesh samples decrease progressively during storage at 30^0c with 70% RH and at 7^0c with 90% RH. The sensory scores of fresh Sandesh sample for flavor, body and texture, colour and appearance were 8.0, 8.0 and 7.5 respectively (Sen and Rajorhia, 1990).

2.3.8: Rheological Study

Rheological properties like hardness, cohesiveness, adhesiveness, chewiness, gumminess and springiness of the Sandesh samples can be described by instrumental means. The texture of Sandesh, was characterized by Khamrui and Solanki, (2010), according to them hardness, fracturability, chewiness and adhesiveness was highest for Sandesh prepared from milk containing 6% fat, 9% SNF and lowest for Sandesh prepared from milk containing 1.5% fat, 9% SNF. Homogenization of buffalo milk intended for mozzarella cheese making led to significant reduction in the hardness, cohesiveness, springiness, gumminess and chewiness of the cheese (Jana and Upadhyay, 1991).

CHAPTER III

MATERIALS AND METHODS

This section deals with the method and materials used in the preparation of Sandesh together with standard analytical procedures.

3.1 Location of Study and Procurement of Milk

The present study was carried out in the Department of Livestock Products Technology, College of Veterinary Science & Animal Husbandry, in Collaboration with Shree G.N.Patel College of Dairy Science & Food Technology, SDAU, Sardarkrushinagar. and Seth M.C. College of Dairy Science, AAU, Anand. Milk (buffalo and cow) were procured from sikaria milk co-operative society, village-sikaria, nearer to Sardar krushinagar, Dantiwada.

3.2 Plan of Work

To accomplish the objectives, the present investigation is being planned with the following mentioned approaches:

The whole study was conducted in two phases. In the first phase, each processing parameter (Table-3.2.1 & 3.2.2) of buffalo milk (compared with control) was selected on the basis of sensory evaluation. Three trials were under taken in case of each parameter. In the second phase, Sandesh made by using all finally selected processing parameters were compared with control (five samples were made for each) on the basis of sensory evaluation. Then the samples were subjected to Chemical, Microbiological and Rheological analysis. Data obtained from the sensory evaluations was subjected to statistical analysis.

3.2.1 Optimization of Processing Parameters for Chhana from Buffalo Milk

No.	Processing Parameter	1	2	Final Selection
3.2.1.1	Level of Fat in Milk.	4%	5%	-
3.2.1.2	Homogenization at 60^0c in Two Stages.	In first stage-1000 psi, second stage-500 psi.	In first stage-1500 psi, second stage-500 psi.	-
3.2.1.3	Addition of Sodium Citrate Prior to its Heating and Coagulation.	0.2%.	0.3%.	-
3.2.1.4	Final Heating Temperature of Milk.	90^0c	95^0c	-
3.2.1.5	Coagulation Temperature. (1% Citric acid)	70^0c	80^0c	-

3.2.2: Optimization of Processing Parameters for Sandesh from Buffalo Milk

No.	Processing parameter	1	2	Final selection
3.2.2.1	Cooking Temperature of Chhana-Sugar (30 % of Chhana) Mixture.	70^0c	80^0c	

3.3 Manufacture of Sandesh

For study on each parameter, buffalo milk Sandesh as well as control was manufactured as per method describe by Roy *et al.*, 2010, which is given below:

3.3.1 Preparation of Chhana

3.3.1.1 Definition of Chhana

According to PFA (1976) rules, chhana is the product obtained from cow or buffalo or a combination thereof by precipitation with lactic acid, citric acid or aged whey. It should not contain more than 70 per cent moisture, and the milk fat content should not be less than 50.0 per cent of the dry matter.

3.3.1.2 Heating and Coagulation of Milk

The milk was filtered and then heated to 90^0c without holding. The temperature of milk was brought down to 70^0c and was coagulated at this temperature using 1 per cent citric acid solution heated to 70^0c. Citric acid was added with continuous agitation till clear whey separated out.

3.3.1.3 Draining

After complete coagulation, the stirring was stopped and the curd was allowed to settle down for few minute. Then the coagulum was strained for expulsion of way. The cloth containing the coagulated solids is than removed, tied up into a bundle without applying pressure and hang up to drain out whey.

3.3.1.4 Mixing and Kneading of Sugar with Chhana

The chhana was collected and grinded to smooth paste. Sugar was added to 30 per cent by weight of chhana for all the experiments during the study. In the study cane sugar in granular form was added to chhana for all the experimental trials.

3. 3.1.5 Cooking of Chhana Sugar Mixture

In the traditional method of Sandesh making, the cooking of chhana-sugar mixture in a iron karahi is done with great skill by the halwais. This is necessary because chhana-sugar mixture is very heat sensitive and hence it has cooked under controlled temperature. Cooking temperature of chhana sugar mixture was 70^0c and cooking time of 20 minutes during experimental trials in order to get the desirable quality of the product. Then Sandesh was cool to 37^0c for 10 minute and stored at 7^0c for further analysis.

3.4 Analysis

The Sandesh samples were analyzed for the following parameters.

3.4.1 Sensory Evaluation

The Sandesh samples were evaluated organolaptically for different quality attributes like flavour (odour and test), body and texture, color and appearance and overall acceptability, by a selected panel of judges comprising of five members. The score card based on classical 9 point hedonic scale was used for evaluation suggested by Amerine *et al.*, 1965.

Table 3.1: Classical 9 Point Hedonic Scale Method

Sr.No.	Hedonic Rating Score	Score
1	Excellent	9
2	Very good	8
3	Good	7
4	Fair	6
5	Nither good nor bad	5
6	Slightly undesirable	4

7	Poor	3
8	Very poor	2
9	Unacceptable	1

3.4.2 Statistical Analysis

The data generated on various parameters was subjected to statistical analysis using two samples T - test (Snedecor and Cochran, 1980).

3.4.3 Chemical Analysis

The chemical analyses of Sandesh obtain from the Sandesh samples and controls were carried out by adopting the followings methods.

3.4.3.1 Determination of Moisture

The moisture content in Sandesh was determined by slandered gravimetric method described in IS: 1479, part II, 1961. Two grams of accurately weighed Sandesh sample was transferred into a clean, dry, silica dish. Than 4 ml of hot distilled water (approximately 65^0c) was mixed with the sample and mixture spread uniformly in the dish with the help of small glass rod. An additional 1 ml hot distilled water was used to wash off the particles of Sandesh from the glass rod into the dish. The dish was then transferred to an oven at $102^0c \pm 1^0c$ and its contents allowed to dry for 4 hrs. The dish was then cooled in desiccators and weighed accurately. The procedure of heating in the oven (for 30 minute), cooling and weighing was repeated until a constant weight was obtained.

3.4.3.2 Determination of Total Solids

For determination of total solid content of Sandesh, the weight of residue left after complete evaporation of the moisture was recorded for the determination of total solids. The total solids were calculated by using formula given below.

$$\text{Total solids (\%)} = \frac{\text{Weight of residue}}{\text{Weight of sample}} \times 100$$

3.4.3.3 Determination of Fat

Fat content of Sandesh samples were determined by Gerber method as per the procedure described by Ladkani and mulay, 1974. Exactly three gram of well mixed Sandesh was weighed and directly taken onto the 10 ml of sulphuric acid (Specific gravity 1.820 to 1.825 at 20^0c) in cheese butyrometer followed by addition of 8 ml of hot (60^0c) distilled water and 1 ml of isoamyl alcohol (Specific gravity 0.815 to 0.825 at 20^0c). The butyrometer was lock stoppered and content were vigorously shaken in horizontal and vertical direction to digest non-fat substances. Liquid level in the butyrometer was brought to calibration by addition of required amount of water. It was centrifuged in Gerber centrifuge machine for 5 minute. Fat column was read after tempering butyrometer for five minute in water bath at 65^0c.

3.4.3.4 Determination of Protein

Protein content of Sandesh samples were determined by following microkjeldahl method as described by AOAC, 1995. Accurately weighed Sandesh sample was (0.5 g) transferred into 800 ml digestion flask. To this 25 ml concentrated sulphuric acid and 10 g digestion mixture consisting of copper sulphate plus potassium sulphate (2:10, w/w) were added and digested over a flame until the content become transparent. The liquid was allowed to cool, diluted with 200 ml distilled water and neutralized with 50% sodium hydroxide. The content was immediately distilled, collecting the distillate in 250 ml conical flask containing 50 ml saturated boric acid added with three drops of mixed indicator (equal volume of saturated solution of methyl red and 0.1% methylene blue solution, both made in 95% v/v ethanol) until all ammonia has passed over in to the boric acid. The distillate was titrated against 0.1 N sulphuric acid solutions.

Total nitrogen was calculated by the formula:

$$\text{Total nitrogen (per cent, w/w)} = 0.14 \times \frac{(A-B)}{W}$$

Total protein (per cent, w/w) = Total nitrogen × 6.38

Where,

A = ml of 0.1 N sulphuric acid required for test sample

B = ml of 0.1 N sulphuric acid required for blank

W = weight in g of sample taken

3.4.3.5 Determination of Ash

Ash content of Sandesh was determined as described in ISI specification (IS: 1479, part II, 1961). In the silica dish approximately 5 g of Sandesh was weighed accurately and then samples were evaporated to dryness at 102 $\pm$ 2 ^{0}c overnight in a hot air oven. The dishes were then transferred to muffle furnace, which was ignited at 600 ^{0}c for 3.5 h to make ash. The contents were cooled in desiccators and weighed. The ash content was calculated as under;

$$\text{Ash (per cent by weight)} = \frac{100 \text{ x } W_1}{W_2}$$

Where,

W_1 = weight of ash in g

W_2 = weight of Sandesh sample taken in g

3.4.3.6 Determination of Carbohydrate

The carbohydrate content of the Sandesh was calculated by difference.

3.4.3.7 Determination of Titratable Acidity

Titaratable acidity of Sandesh samples were determined by following method as described by Sen and Rajorhia, 1986.

Two gram of Sandesh was accurately weighed in a porcelain dish and rendered into a fine paste with 3.0 ml of hot distilled water, and diluted by another 17.0 ml of hot distilled water, washing off the adherents from the pestle. 10.0 ml of 0.1 N NAOH was added to dissolve the contents and subsequently the excess alkali was titrated against 0.1 N HCL in the presence of 1.0 ml of 0.5% phenolphthalein indicator solution until pink colour disappeared. Acidity was expressed as lactic acid per 100 g of Sandesh using the following formula.

Per cent lactic acidity = $\frac{10-V}{W} \times 0.9$

Where, V= ml of 0.1 N HCL needed for titration.

W=weight of Sandesh taken

3.4.4 Microbiological Analysis

The Microbial analysis (the total viable, coliform and yeast and mould counts) of the Sandesh samples were enumerated using followings methods.

3.4.4.1 Standard plate count

The total numbers of viable bacteria in Sandesh were enumerated by the method described in IS: 1479 (Part III)-1977. Plate count agar obtained from the Hindustan Dehydrated Media (Hi-Media), Bombay was used as a nutrient medium. To rehydrate the medium, 23.5 g were suspended in 1000 ml distilled water and boiled to dissolve the medium completely. It was then filled in flasks and sterilized by autoclaving at 15 psi (121^0c) for 15 min. Appropriate dilutions were prepared from each sample of Sandesh using sterile normal saline solution. One ml quantity each of suitable dilution was placed in duplicate on the nutrient agar medium. The plates were incubated at 37^0c for 48 hrs.

3.4.4.2 Yeast and Mould Count

Yeast and Mould count was done according to the procedure recommended by IS: 5253-1969. Potato dextrose agar (Hi-Media) was used to determine the yeast and mould counts in Sandesh. To rehydrate the medium, 39 g were suspended in 1000 ml distilled water and boiled to dissolve the medium completely. It was then filled in flasks and sterilized by autoclaving at 15 psi (121^0c) for 15 min.

Appropriate dilutions were prepared from each sample of Sandesh using sterile normal saline solution. One ml quantity each of suitable dilution was placed in duplicate on Potato dextrose agar medium by adjusting temperature of pouring to 45^0c. The pH of the medium was adjusted to 3.5 at the time of plating by using sterile 10 per cent tartaric acid. The plates were incubated at 22^0c for 3-5 days and growth was noted.

3.4.4.3 Coliform counts

Coliform counts was done according to the procedure specified by the method explained in IS: 1479 (Part III)-1977. Violet red bile agar (Hi-Media) was used for enumeration of coliform in Sandesh. To rehydrate the medium, 41.5 g were suspended in 1000 ml distilled water and boiled to dissolve the medium completely. The medium was not autoclaved. Appropriate dilutions were prepared from each sample of Sandesh using sterile normal saline solution. One ml quantity each of suitable dilution was placed in duplicate on Violet red bile agar. The plates were incubated at 37^0c for 24 hrs.

3.4.5 Rheological Analysis

Various textural characteristics of Sandesh viz. hardness, cohesiveness, adhesiveness, chewiness, gumminess and springiness of the Sandesh samples were measured from force –time curve (Bourne, 2002). Cuboids samples of the product measuring 20 mm each in length and width and 15 mm high and five samples of each experimental Sandesh were subjected to uniaxial compression to 70% of the initial sample height, using a Food Texture Analyzer of Lloyd Instruments LRX plus Material Testing Machine, England; fitted with 0-500 kg load cell. The force-distance curve was obtained for a two-bite deformation cycle employing a cross head speed of 50 mm/min, trigger 10 gf and 70% compression of the samples to determine various textural attributes of Sandesh held for 1 h at 23 ± 1^0C and 55% RH. The complete work of calculations of area under the force-distance curve. The statistical analysis of data obtain from Rheological study was subjected to statistical analysis using two sample T - test (Snedecor and Cochran, 1980).

The following textural parameters were interpreted from the graph as under:

Hardness (N):

The force necessary to attain a given deformation, i.e. the highest point of the peak in the first bite curve, at 70.0 per cent compression.

Cohesiveness:

The extent to which a material can be deformed before it ruptures.

$$\text{Cohesiveness} = \frac{A2}{A1}$$

Where, A_1 = Area under the first bite curve before reversal of compression.

A_2 = Area under the second bite curve before reversal of compression.

Springiness (mm):

It is the height that the sample recovers between the first and second compression on removal of the deforming forces.

Gumminess (N):

It is the energy required to masticate a sample to a state ready for swallowing. It is a product of hardness (N) and cohesiveness multiplied by 100.

$$\text{Gumminess} = \text{hardness} \times \text{cohesiveness} \times 100$$

Chewiness (Nmm):

The energy required to masticate a food to a state ready for swallowing. It is a product of hardness (N), cohesiveness and springiness (mm)

$$\text{Chewiness} = \text{hardness} \times \text{cohesiveness} \times \text{springiness}$$

Adhesiveness (Nmm):

It is the work necessary to overcome the attractive forces between the surfaces of the sample and the other materials with which the sample comes in contact. It is the negative force area for the first bite curve.

CHAPTER IV

RESULTS

The whole study was conducted in mainly two phases. In the first phase each processing parameter of buffalo milk (compared with control) was selected on the basis of sensory evaluation. Three trials were taken in case of each parameter. In the second phase Sandesh made by using all finally selected processing parameters were compared with control (five samples were made for each) on the basis of sensory evaluation. Then the samples were subjected to chemical, microbiological and rheological analysis. Data obtained from the sensory evaluation was subjected to statistical analysis.

4.1: Effects of Fat Level

It was observed that Sandesh prepared using 4% and 5% buffalo milk resulted in hard body and coarse texture as compare to control. Brittleness was also found in both the samples (2 & 3).The score of colour & appearance was also lower than control. Among the two type of buffalo milk Sandesh, the score of all sensory attributes of samples-2 was more than that of samples-3. There was minute amount of fat leakage in case of samples-2. The score of all samples are significantly different at both 5% and 1% level of significance. Sensory characteristic of Sandesh made from buffalo milk standardized at different fat levels are presented in Table-4.1 (a). Five per cent fat was selected for manufacture of final product.

Table 4.1(a): Effects of Fat Level

Samples	Flavour	Body & Texture	Colour & Appearance	Overall acceptability
Samples-1	7.93	8.03	7.95	7.92
Samples-2	6.9	6.27	6.58	6.48
Samples-3	6.53	5.6	6.15	5.9

Samples-1: control (Sandesh made from cow milk), **Samples-2:** Sandesh made from Buffalo milk having 5% fat, **Samples-3:** Sandesh made from Buffalo milk having 4% fat.

Table 4.1(b): T- Test for Effects of Fat Percentage

Test of significant in-between sample-1 & 2, sample-1 & 3 and sample-3 & 2

T- test	Sample-1	Sample-2	Sample-3
Number of Observation	15	15	15
Average	31.833	26.233	24.183
Standard Deviation	0.323	0.406	0.671
Variance	0.104	0.165	0.451
Test results	**A****	**B****	**C** **
T – Statistic	41.814	39.779	10.120
T – Table (0.05)	2.048	2.145	2.145
T – Table (0.01)	2.763	2.977	2.977

** Samples are significantly different at both 5% and 1% level of significance.

A: Test results in-between sample-1 and 2, **B**: Test results in-between sample-1 and 3. **C**: Test results in-between sample-2 and 3

Fig. 4.1: Photograph showing effects of fat level on buffalo milk Sandesh

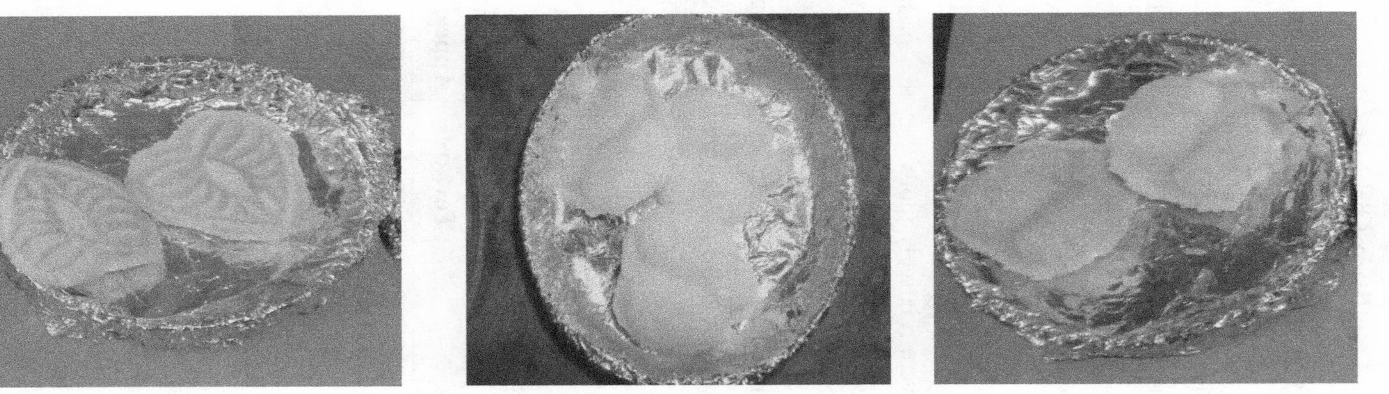

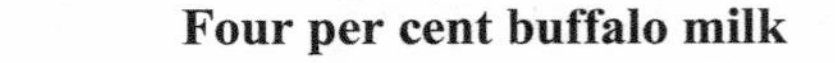

Control | **Five per cent buffalo milk** | **Four per cent buffalo milk**

4.2: Effects of Sodium Citrate

Since buffalo milk Sandesh resulted in hard, coarse and brittle texture, attempt was made to prepare by addition of sodium citrate to different level (0.2% and 0.3%) prior to heating and coagulation of milk. The trials showed that there was major enhancement in body and texture of both buffalo milk Sandesh (samplse-2 and samples-3), though there was some problem of fat leakage. The score of flavour, body, texture, colour & appearance and overall acceptability of samples-2 and samples-3 was lower than that of control, whereas in between the two types of buffalo milk Sandesh, the score of samples-3 were relatively more than that of samples-2. It is revealed that score of Sandesh made from buffalo milk added with different levels (0.3% and 0.2%) of sodium citrate are significantly different at 5% level of significance but not at 1%. The results pertaining to sensory characteristic of buffalo milk Sandesh as influenced by levels of sodium citrate are depicted in Table-4.2 (a). Addition of sodium citrate to 0.3% prior to heating and coagulation of milk was selected for manufacture of final product.

Table 4.2(a): Effects of Sodium Citrate

Samples	Flavour	Body & Texture	Colour & Appearance	Overall acceptability
Samples-1	8.07	8.15	8.02	8.05
Samples-2	7.42	7.28	7.25	7.43
Samples-3	7.43	7.52	7.43	7.5

Samples-1: control (Sandesh made from cow milk), **Samples-2:** sodium citrate added to 0.2%, **Samples-3:** Sodium citrate added to 0.3%

Table 4.2(b): T- Test for Effects of Sodium Citrate

Test of significant in-between sample-1 & 2, sample-1 & 3 and sample-3 & 2

T- test	**Sample-1**	**Sample-2**	**Sample-3**
Number of Observation	15	15	15
Average	32.283	29.383	29.883
Standard Deviation	0.339	0.449	0.706
Variance	0.115	0.249	0.499
Test results	**A****	**B****	**C** *
T – Statistic	18.624	11.865	2.240
T – Table (0.05)	2.048	2.145	2.048
T – Table (0.01)	2.763	2.977	2.763

* Samples are significantly different at 5% level of significance.

** Samples are significantly different at both 5% and 1% level of significance.

A: Test results in-between sample-1 and 2,

B: Test results in-between sample-1 and 3,

C: Test results in-between sample-3 and 2

Fig. 4.2: Photograph showing effects of sodium citrate on buffalo milk Sandesh

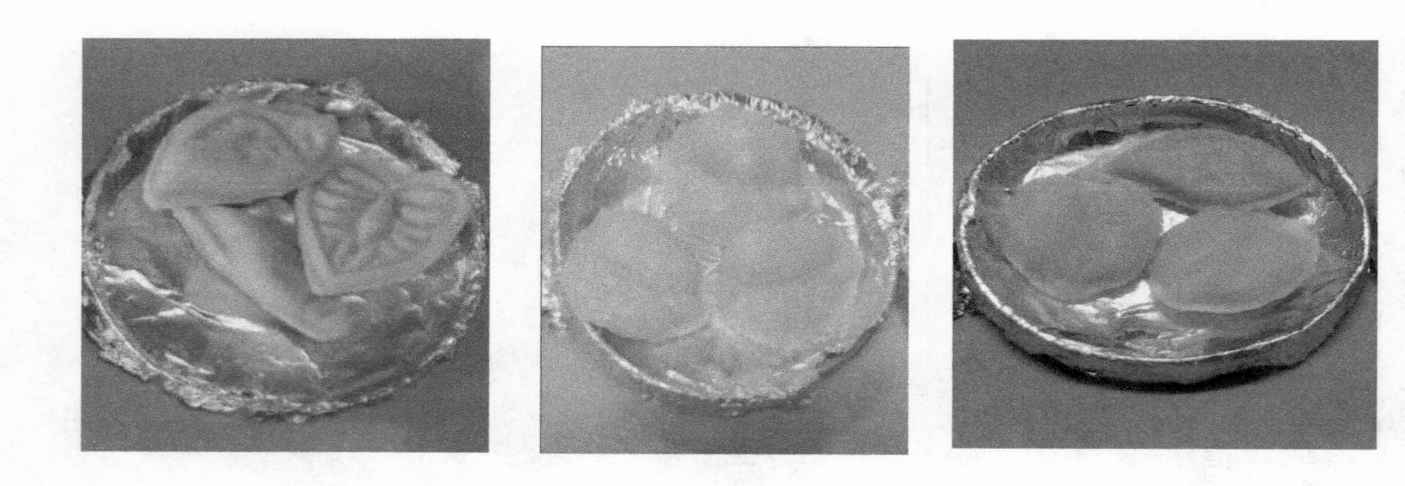

Control | **Addition of sodium citrate at 0.3%** | **Addition of sodium citrate at 0.2%**

4.3: Effects of Homogenization

Milk was homogenized in two stages at 60°c. The pressures in first stage were 1500 psi and 1000 psi in case of samples-2 & 3 respectively, whereas the pressures in second stage were 500 psi. in case of both. It was observed that samples-2 resulted in better body and texture without leakage of fat, whereas samples-3 resulted in comparatively hard body, coarse texture and poor in overall acceptability. The samples made from milk homogenized at1500 psi were close to control. The flavour of the Sandesh samples made from buffalo milk homogenized at 1500 psi. were similar to slightly better than control, whereas the colour and appearance of the samples were slightly inferior to control. The sensory score of control and Sandesh samples made from buffalo milk homogenized at1500 psi are significantly different at 5% level of significance but not at 1%. The data concern to sensory attribute of samples-1, 2 and 3 are presented in Table-4.3 (a). The pressure in first stage at1500 psi and 500 psi in second stage for homogenization of buffalo milk was selected for manufacture of final product.

Table 4.3(a): Effects of Homogenization

Samples	Flavour	Body & Texture	Colour & Appearance	Overall acceptability
Samples-1	7.81	8.02	7.93	7.93
Samples-2	7.88	7.92	7.73	7.82
Samples-3	7.2	6.43	6.7	6.63

Samples-1: control (Sandesh made from cow milk), **Samples-2:** Homogenization of milk at 1500 psi, **Samples-3**: Homogenization of milk at 1000 psi.

Table 4.3(b): T- Test for Effects of Homogenization

Test of significant in-between sample-1 & 2, sample-1 & 3 and sample-3 & 2

T- test	**Sample-1**	**Sample-2**	**Sample-3**
Number of Observation	15	15	15
Average	31.783	31.350	26.983
Standard Deviation	0.499	0.461	0.571
Variance	0.249	0.213	0.326
Test results	**A***	**B****	**C** **
T – Statistic	2.241	24.529	23.055
T – Table (0.05)	2.048	2.048	2.048
T – Table (0.01)	2.763	2.763	2.763

* Samples are significantly different at 5% level of significance.

** Samples are significantly different at both 5% and 1% level of significance.

A: Test results in-between sample-1 and 2,

B: Test results in-between sample-1 and 3,

C: Test results in-between sample-2 and 3

Fig. 4.3:Photograph showing effects of homogenization on buffalo milk Sandesh

Control | **Homogenization at1000 psi** | **Homogenization at1500 psi**

4.4: Effects of Final Heating Temperature

In order to enhance the acceptability, the milk was heated to 90°c in case of samples-2, whereas in case of samples-3 milk was heated to 95°c. Thereafter the chhana was made by coagulating milk with 1% citric acid solution. Out of the two final heating temperatures, 90°c result in a product with somewhat better overall acceptability than product at 95 °c. The sensory score of Sandesh samples made from buffalo milk chhana at 90°c final heating temperatures and at 95 °c final heating temperatures are significantly different at 5% level of significance but not at 1%. The score of both buffalo milk Sandesh samples were lower than that of control. The sensory score of control and both buffalo milk Sandesh samples are significantly different at both 5% and 1% level of significance. The data regarding the sensory characteristic of Sandesh samples made from buffalo milk chhana at various final heating temperatures are presented in Table-4.4 (a). Final heating temperature of milk at 90°c was selected for manufacture of final product.

Table 4.4(a): Effects of Final Heating Temperature

Samples	**Flavour**	**Body & Texture**	**Colour & Appearance**	**Overall acceptability**
Samples-1	7.87	8.02	7.9	7.93
Samples-2	6.82	5.83	6.12	6.12
Samples-3	6.53	5.63	6.15	6.0

Samples-1: control (Sandesh made from cow milk), Samples-**2:** Final heating temperature for chhana preparation at 90°c, **Samples-3:** Final heating temperature for chhana preparation at 95°c.

Table 4.4(b): T- Test for Effects of Final Heating Temperatures

Test of significant in-between sample-1 & 2, sample-1 & 3 and sample-3 & 2

T- test	**Sample-1**	**Sample-2**	**Sample-3**
Number of Observation	15	15	15
Average	31.717	24.867	24.317
Standard Deviation	0.432	0.364	0.704
Variance	0.186	0.133	0.495
Test results	**A****	**B****	**C ***
T – Statistic	46.969	34.716	2.688
T – Table (0.05)	2.048	2.145	2.145
T – Table (0.01)	2.763	2.977	2.977

* Samples are significantly different at 5% level of significance.

** Samples are significantly different at both 5% and 1% level of significance.

A: Test results in-between sample-1 and 2,

B: Test results in-between sample-1 and 3,

C: Test results in-between sample-2 and 3

Fig. 4.4: Photograph showing effects of final heating temperature on buffalo milk Sandesh

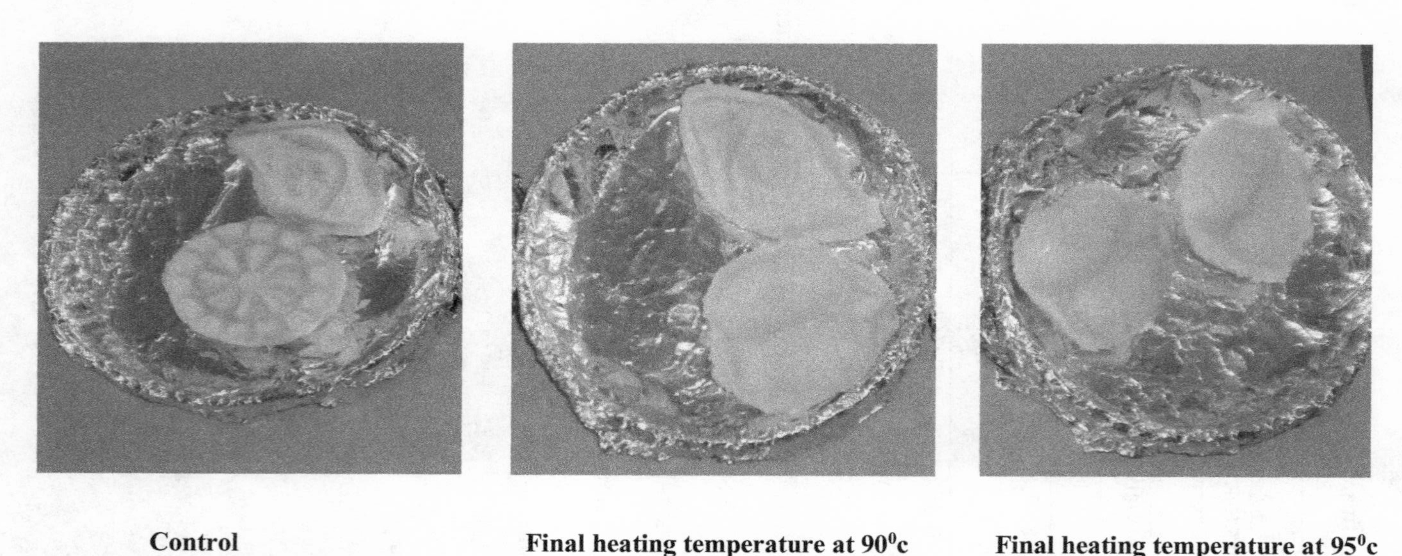

Control **Final heating temperature at 90^0c** **Final heating temperature at 95^0c**

4.5: Effects of Coagulation Temperature

To study the effect of different coagulation temperatures, Buffalo milk was heated to 90°c without any holding followed by coagulation at 80 and 70 °c by 1% citric acid solution. The effect of temperature of coagulation was more conspicuous on body and texture of Sandesh. As the coagulation temperature increase, the body becomes harder. There was a little marked affect on the flavour and colour and appearance of the product due to temperature of coagulation. The yield of chhana decreased as coagulation temperature increased. The yield of chhana was highest (24.7%) from buffalo milk coagulated at 70 °c followed by coagulated at 80 °c (21.9%) and control coagulated at 70 °c (17.8%). The score of samples-2 and 3 was poor than that of control but among the two buffalo milk Sandesh, the score of Sandesh samples made from buffalo milk coagulated at 70 °c was better than coagulated at 80 °c. The score of all samples are significantly different at both 5% and 1% level of significance. The effects of different coagulation temperature on sensory attributes of different Sandesh samples are shown in Table-4.5 (a). Coagulation temperature of milk at 70°c was selected for manufacture of final product.

Table 4.5(a): Effects of Coagulation Temperature

Samples	**Flavour**	**Body & Texture**	**Colour & Appearance**	**Overall acceptability**
Samples-1	8.02	8.12	7.98	8.0
Samples-2	6.82	5.72	5.82	5.93
Samples-3	6.62	5.33	5.5	5.65

Samples-1: control (Sandesh made from cow milk), **Samples-2:** Coagulation temperature of milk at 70 °c for chhana preparation, **Samples-3:** Coagulation temperature of milk at 80 °c for chhana preparation.

Table 4.5(b): T- Test for Effects of Coagulation Temperature

Test of significant in-between sample-1 & 2, sample-1 & 3 and sample-3 & 2

T- test	**Sample-1**	**Sample-2**	**Sample-3**
Number of Observation	15	15	15
Average	32.150	24.283	23.100
Standard Deviation	0.351	0.640	0.604
Variance	0.123	0.410	0.364
Test results	**A****	**B****	**C** **
T – Statistic	41.743	50.200	5.210
T – Table (0.05)	2.145	2.145	2.048
T – Table (0.01)	2.977	2.977	2.763

** Samples are significantly different at both 5% and 1% level of significance.

A: Test results in-between sample-1 and 2,

B: Test results in-between sample-1 and 3,

C: Test results in-between sample-2 and 3

Fig. 4.5: Photograph showing effects of coagulation temperature on buffalo milk Sandesh

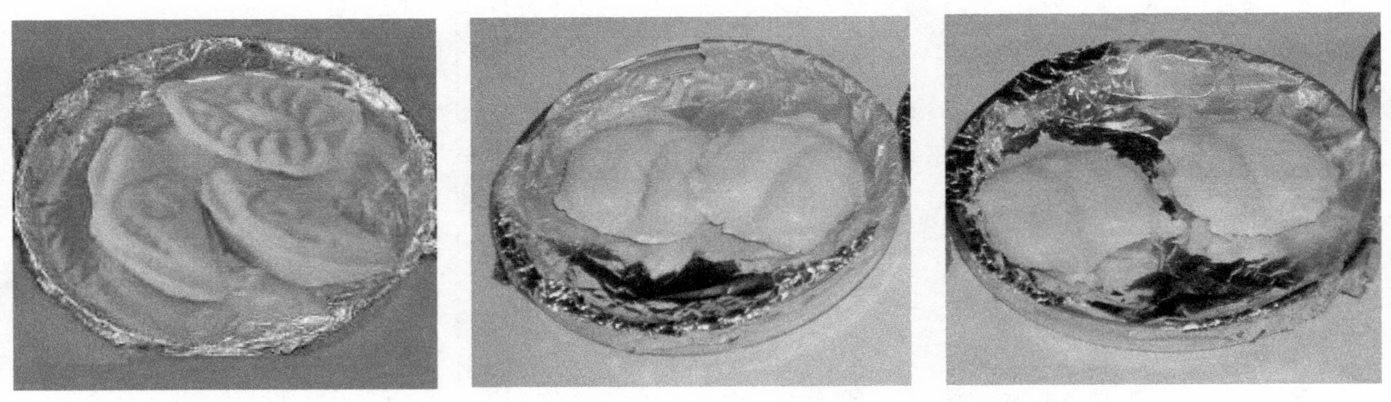

Control **Coagulation temperature at 70^0c** **Coagulation temperature at 80^0c**

4.6: Effects of Cooking Temperature

To study the effect of different cooking temperature on Sandesh, chhana- sugar mixture was cooked at 70 °c in case of samples-2, whereas in case of samples-3 mixture was cooked at 80°c. The mixture which was cooked at 70 °c for 20 minute resulted in comparatively soft Sandesh, whereas the mixture which was cooked at 80 °c for 20 minute gave slightly hard, brittle and chewy product. The score of samples- 2 was more in all aspect than that of samples- 3, while the score of these buffalo milk Sandesh was poor than that of control. The sensory score of control and buffalo milk Sandesh samples are significantly different at both 5% and 1% level of significance. The effects of different cooking temperature on sensory attributes of different Sandesh samples are shown in Table-4.6 (a). Cooking temperature of chhana-sugar mixture at70 °c for 20 minute was selected for manufacture of final product.

Table 4.6(a): Effects of Cooking Temperature

Samples	Flavour	Body & Texture	Colour & Appearance	Overall acceptability
Samples-1	7.92	8.07	7.82	7.93
Samples-2	6.48	5.67	5.83	5.87
Samples-3	6.2	5.43	5.5	5.72

Samples-1: control (Sandesh made from cow milk), **Samples-2:** Cooking temperature for sandesh at 70 °c, **Samples-3:** Cooking temperature for sandesh at 80 °c.

Table 4.6(b): T- Test for Effects of Cooking Temperature

Test of significant in-between sample-1 & 2, sample-1 & 3 and sample-3 & 2

T- test	**Sample-1**	**Sample-2**	**Sample-3**
Number of Observation	15	15	15
Average	31.767	23.850	22.850
Standard Deviation	0.395	0.712	0.646
Variance	0.156	0.507	0.418
Test results	**A****	**B****	**C** **
T – Statistic	37.653	45.589	4.027
T – Table (0.05)	2.145	2.145	2.048
T – Table (0.01)	2.977	2.977	2.763

** Samples are significantly different at both 5% and 1% level of significance.

A: Test results in-between sample-1 and 2,

B: Test results in-between sample-1 and 3,

C: Test results in-between sample-2 and 3

Fig. 4.6: Photograph showing effects of cooking temperature on buffalo milk Sandesh

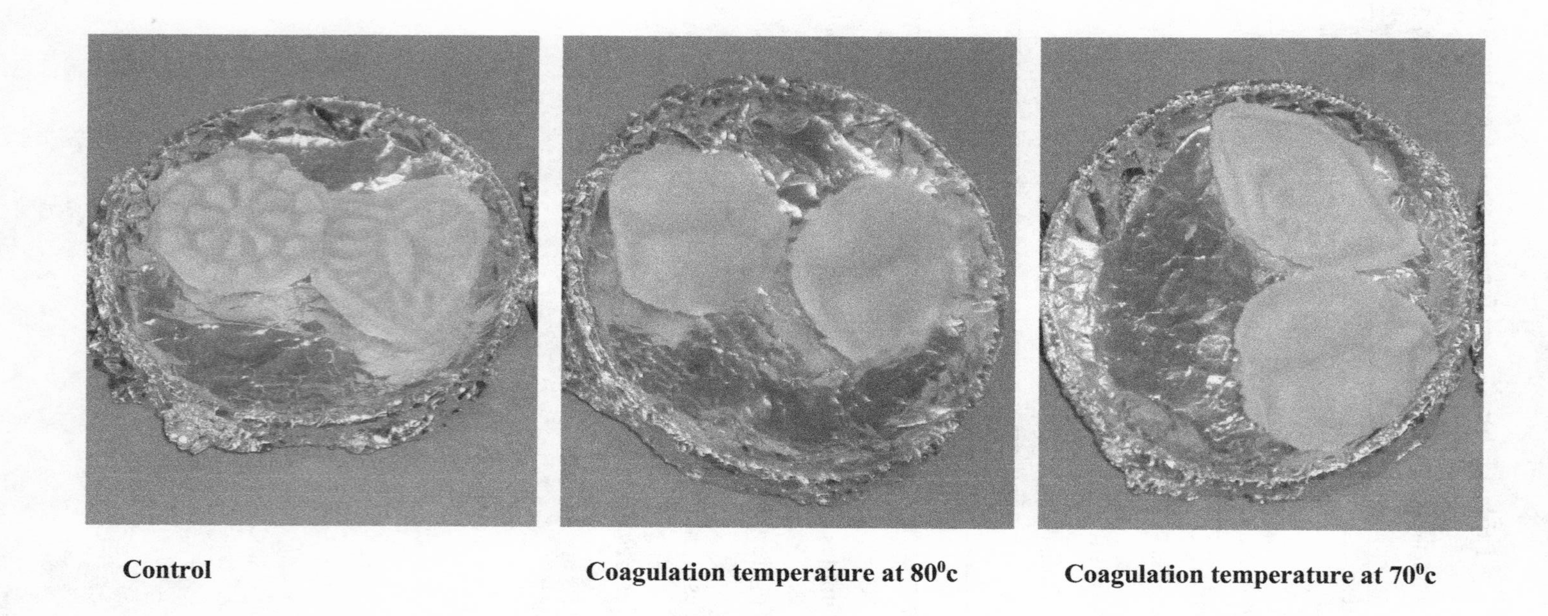

Control | **Coagulation temperature at 80^0c** | **Coagulation temperature at 70^0c**

4.7: Evaluation and Analysis of Final Product

The final product was manufactured from buffalo milk using all selected processing parameter from previous experiments (5% fat level, 0.3% sodium citrate, homogenization at1500 psi, final heating temperature at 90°c, coagulation temperature at 70°c and cooking temperature70°c for 20 minute). The sensory score for flavour, body and texture of final product was slightly more in compare to that of control whereas score for colour and appearance was comparatively less to that of control. There was no fat oozing on the surface of final product. The body and texture of final product was further improved than that of Sandesh made from previous individual experiments with homogenization and sodium citrate. The sensory score of samples of final product and control are significantly different at 5% level of significance but not at 1%. The effects all finally selected parameter on sensory features of buffalo milk Sandesh samples are shown in Table-4.7 (a).

Table 4.7(a): Sensory Evaluation of Final Product

Samples	Flavour	Body & Texture	Colour & Appearance	Overall acceptability
Samples-1	7.92	7.98	7.98	7.92
Samples-2	8.08	8.2	7.78	8.07

Samples-1: control (Sandesh made from cow milk).

Samples-2: Sandesh made from buffalo milk using all selected processing parameters.

Table 4.7(b): T- Test for Sensory evaluation of final product

Test of significant in-between sample-1 & 2, sample-1 & 3 and sample-3 & 2

T- test	**Sample-2**	**Sample-1**
Number of Observation	15	15
Average	32.133	31.817
Standard Deviation	0.376	0.334
Variance	0.142	0.111
Test results	**A***	
T – Statistic	2.438	
T – Table (0.05)	2.048	
T – Table (0.01)	2.763	

*Samples are significantly different at 5% level of significance.

A: Test results in-between sample-2 and 1.

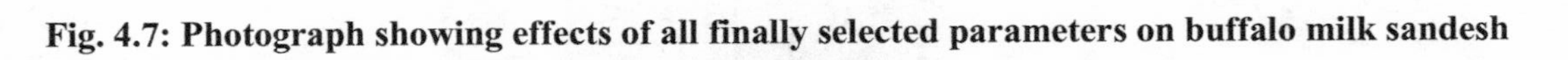
Fig. 4.7: Photograph showing effects of all finally selected parameters on buffalo milk sandesh

4.7.1: Chemical Composition

It is clear from the data that there was some variation in the chemical composition of control and experimental Sandesh (final product). All the samples showed variation in moisture content of the final product. Sandesh made from control recorded more total solid (74.87%, 75.44% and 75.58%) as compare to that of buffalo milk Sandesh (73.58%, 74.14% and 73.77%). The fat and carbohydrate content decreased but the protein and ash contents increase slightly in the product from buffalo milk Sandesh. The titaratable acidity of control was somewhat more to that of buffalo milk Sandesh. The major chemical composition of control and final product are given in Table-4.8.

Table 4.8: Chemical Composition of Control and Buffalo Milk Sandesh (Final Product)

Constituent	Composition (%)							
	C_1	C_2	C_3	Average	B_1	B_2	B_3	Average
Moisture	25.13	24.56	24.42	24.70	26.42	25.86	26.33	26.20
Total solids	74.87	75.44	75.58	75.29	73.58	74.14	73.77	73.83
Fat	18.96	19.22	19.56	19.24	18.52	19.12	18.23	18.62
Protein	18.42	18.28	18.44	18.38	18.71	18.92	19.14	18.92
CHO	34.98	35.47	35.00	35.15	33.63	33.34	33.55	33.50
Ash	1.62	1.55	1.68	1.61	1.86	1.88	1.90	1.88
Titerable acidity	0.89	0.92	0.90	0.90	0.86	0.88	0.85	0.86

4.7.2: Microbial Study

The average total viable, yeast and mould and Coliform counts/g of Sandesh made from buffalo milk were17.7 $x10^3$, 5 x $10^1$and $2x10^1$ respectively. The microbial counts of Sandesh samples are given in Table-4.9.

Table 4.9: Microbial Counts of Control and Buffalo Milk Sandesh (Final Product)

Samples	Organisms Counts/g		
	Total Viable Count	**Yeast-Mold Count**	**Coliform Count**
B_1	4×10^4	1×10^1	2×10^1
B_2	5×10^3	6×10^1	3×10^1
B_3	8×10^3	8×10^1	1×10^1
Average	17.7×10^3	5×10^1	2×10^1

Where,

C= Control B= Buffalo milk Sandesh (Final Product)

4.7.3 Rheological Properties of Sandesh

The average value of hardness, cohesiveness, springiness, gumminess, chewiness and adhesiveness of control were 132.546 N, 0.038, 1.464 mm, 4.496 N, 7.082 Nmm and 0.817 Nmm while for final product these values were 118.188 N, 0.017, 0.989 mm, 1.177 N, 1.651 Nmm and 0.183 Nmm respectively. The numerical values of texture profile parameters are presented in Table-4.10.

Table 4.10: Rheological values of control and Buffalo milk Sandesh (Final product)

Rheological Properties	Samples									
	C 1	C2	C3	C 4	C 5	B1	B2	B3	B4	B5
Sample Height (mm)	15	15	15	15	15	15	15	15	15	15
Hardness1 (N)	131.468	129.908	147.339	125.379	128.639	100.905	116.573	121.087	125.614	126.735
Hardness2 (N)	61.9253	64.1153	70.0054	59.2739	62.3701	50.6978	54.3904	58.7434	60.4721	60.4345
Cohesiveness	0.04109	0.05605	0.02988	0.01199	0.03027	0.00857	0.01677	0.01641	0.02067	0.01148
Springiness (mm)	1.72262	1.71831	1.18352	0.85501	1.82803	1.15930	0.83089	1.12208	0.94734	0.84499
Gumminess (N)	5.40202	7.281343	4.40248	1.50372	3.89390	0.86492	1.95492	1.98703	2.59644	1.45491
Chewiness (Nmm)	9.30568	12.51161	5.21043	1.2857	7.11817	1.0028	1.62433	2.22962	2.45971	1.22939
Adhesiveness (Nmm)	0.74366	0.556571	0.82728	0.18203	1.76908	0.16325	0.19994	0.16286	0.19590	0.21457

Where, C= Control B= Buffalo milk Sandesh (Final product)

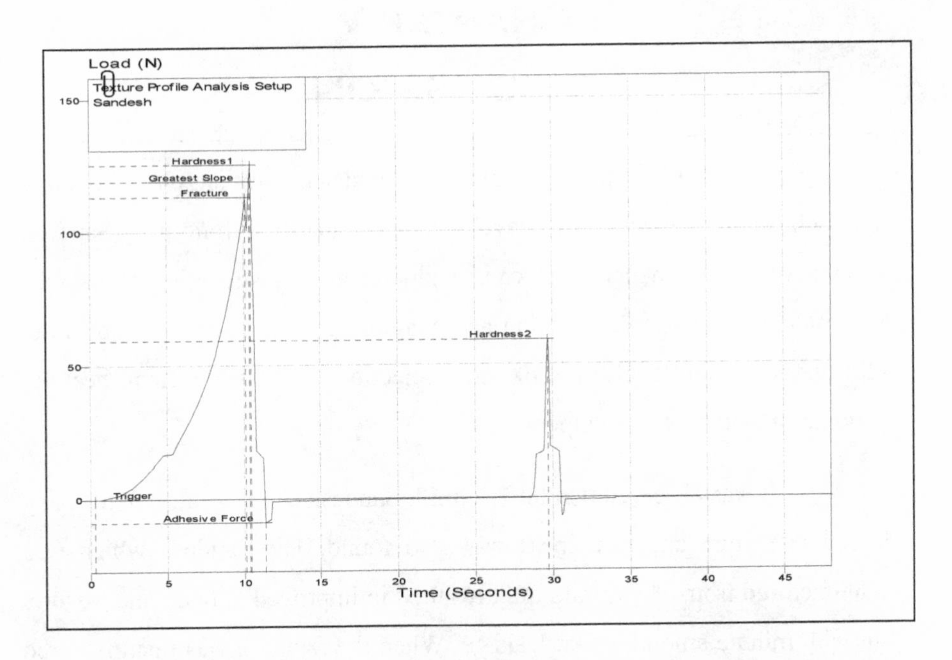

Fig. 4.8: A Typical Texture Profile Analysis Curve of Control

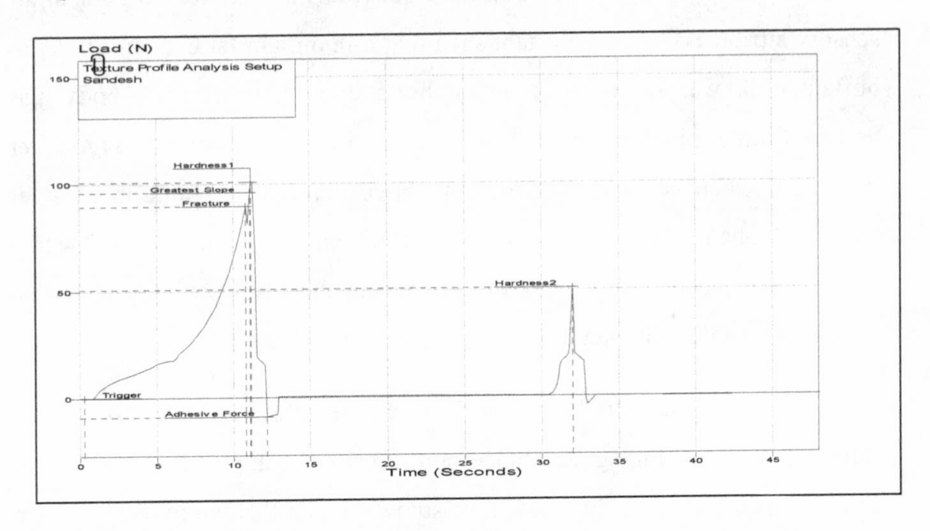

Fig. 4.9: A Typical Texture Profile Analysis Curve of Final Product

CHAPTER V
DISCUSSION

The present study has clearly elucidated how simple innovative approaches could be applied to developed good quality buffalo milk Sandesh by some of the treatments which can be adopted are adjustment in salt balance of buffalo milk prior to its heating and coagulation, homogenization of milk, adjustment of fat level in milk and selection of suitable final heating, coagulation and cooking temperatures.

It was observed that buffalo milk Sandesh could result in hard body & coarse texture and brittleness was also found. The Sandesh which was manufactured from 5% buffalo milk resulted in improved in body and texture, but with minute amount of fat leakage. When the Sandesh was manufactured from 4 % buffalo milk, resulted in hard body, coarse texture and poor in all sensory attributes. It is well established that a minimum fat content of 5% in buffalo milk was essential to obtain Sandesh with satisfactory body and texture. Similar observation was made by De (1980). According to Jagtiani *et al.,* 1960, when calcium was present above a particular limit in milk, it rendered chhana hard. This could be due to curd tension of milk, which influences the body and texture of chhana and thereby hard body & coarse texture of buffalo milk Sandesh.

When 0.25 to 0.30% sodium citrate added to milk to replace calcium, which in turn gave reduce curd tension to improve the quality of chhana made from buffalo milk. Our study showed that addition of sodium citrate into buffalo milk at 0.2% and 0.3% prior to its heating and coagulation were resulted in soft body and smooth texture, though the product was still not satisfactory due to problem of fat leakage. The Sandesh manufactured from milk added with sodium citrate at 0.3% was comparatively more acceptable than that of 0.2%. Our observation is in close agreement with those reported by Jagtiani *et al.* (1960). The beneficial effect of sodium citrate in the manufacture of Sandesh from buffalo milk was also reported by De and Ray, 1954. However, the liking of these Sandesh, were comparatively less than the control made from cow milk. This may be due to leakage of fat into product and comparatively less acceptable than control on the basis of sensory evaluation (flavour, body and texture and colour and appearance).

It is apparent from the previous trials that addition of sodium citrate prior to its heating and coagulation had tendency to improve in body and texture, though the product was still not satisfactory due to problem of fat leakage. This gave clue to do homogenization of milk which would ensure softer Sandesh without leakage of fat. As stated by Kanawjia, (1975), a homogenization pressure of 1900 psi should be adequate to improve the body and texture of buffalo milk chhana. Our study explained that if milk was homogenized in two stages in which the pressures in first and second stage were 1500 and 500 psi respectively resulted in better body and texture without fat leakage and this product was close to control. This could be due to subdivision of fat globules and their interaction with casein micelles resulting in a compaction effect. In contrast, the milk which was homogenized at lower pressure (1000 and 500 psi) gave poor body and texture.

Sandesh was manufactured with different final heating temperature (90°c and 95°c). The study demonstrated that Sandesh manufactured at 90°c final heating temperature was very close to the Sandesh manufactured at 95°c final heating temperature, though overall acceptability was relatively more in case of Sandesh manufactured at 90°c than that of 95°c. This observation is in close agreement with those reported by Sahu and Jha, (2008), According to them for good quality chhana for Sandesh preparation, milk was heated to 90^0c and then coagulated at 70^0c.

As reported by Davies, (1948), coagulation of milk in chhana making is due to combined effect of chemical and physical changes in the casein micelles brought about by the action of acid aided by relatively higher temperature of coagulation. Our study proved that coagulation of milk at 80 °c resulted in low yield and comparatively hard body and coarse texture whereas coagulated at 70 °c the yield was more and body and texture was comparatively desirable. The improvement in body and texture could be due to the reason that as coagulation temperature decreases, the moisture retention in chhana increases leading to its softer body and smooth texture (Sen., 1986). Our observation is in close agreement with those accounted by Kundu and De, (1972), Gajendra, (1976), Iyer, (1978), Soni *et al*., 1980 and Ahmed *et al.,* 1981 that the optimum coagulation temperature for chhana making from buffalo milk is 70°c.

The main process of Sandesh preparation is the cooking of chhana sugar mixture to reduce its moisture and to obtain the characteristic flavour and texture. Our study showed that various cooking temperature had significant effect on quality of Sandesh. When the mixture was cooked at 70°c for 20 minutes resulted in relatively soft Sandesh having comparatively better body, texture & overall acceptability and when the mixture was cooked at 80°c for 20 minutes gave slightly hard, brittle and chewy product. According to Das, 2000 and Sen and Rajorhia, 1985, in the traditional method of Sandesh manufacture, the chhana sugar mixture is cooked in a karahi kept directly over fire at 75 to 85^0c for 15 to 25 minutes depending upon type of

Sandesh. The cooking at higher (>85) temperature gave hard, brittle and chewy product with little cohesiveness.

The final product was manufactured from buffalo milk using all selected processing parameter. The study demonstrated that there was a further improved in flavour, body and texture of final product than that of Sandesh made from previous individual experiments with homogenization and sodium citrate. These improvements might be mainly due to combine effect of homogenization and sodium citrate. There was no free fat on the surface of final product lead to devoid of fat leakage. This may be due to homogenization. In this study, it was also observed that flavour, body and texture of final product was fairly enhanced than that of control, though colour and appearance was slightly lower to that of control.

According to Verma, (1997), there was some variation in the chemical composition of cow milk Sandesh and buffalo milk Sandesh. Our study exposed that all the samples showed variation in moisture content. Sandesh made from control recorded more total solid (74.87%, 75.44% and 75.58%) as compare to that of buffalo milk Sandesh (73.58%, 74.14% and 73.77%). This was due to lower initial moisture content in chhana resulting into lower moisture lead to higher total solid content in the final product. The fat and carbohydrate content was low in buffalo milk Sandesh than that of control, it might be due to comparatively high moisture content as compare to control. In contrast the protein and ash content was more. Further, there might no effect of homogenization and sodium citrate on the chemical composition of products. Our findings of chemical composition are close to chemical composition of cow milk Sandesh and buffalo milk Sandesh reported by Verma, 1997. The average total viable, yeast and mould and coliform counts/g of final products were17.7 $x10^3$, 5 x $10^1$and $2x10^1$ respectively.

The quality of product is monitored not only by the sensory properties but also by their rheological/texture profile. Our study showed that there were significant reduction in the hardness, cohesiveness, springiness, gumminess and chewiness of the final product as compare to that of cow milk Sandesh. It might be due to homogenization and addition of sodium citrate in to buffalo milk prior to heating and coagulation. Similar report on mozzarella cheese given by Jana and Upadhyay (1990), According to them homogenization of buffalo milk intended for mozzarella cheese making led to significant reduction in the hardness, cohesiveness, springiness, gumminess and chewiness of the cheese.

CHAPTER VI
SUMMARY AND CONCLUSIONS

This investigation mainly covers the effect of various processing parameters like fat percentage of milk, addition of sodium citrate in milk prior to its heating and coagulation, homogenization of milk, final heating temperature, coagulation temperature and cooking temperature for the manufacture of Sandesh from buffalo milk. It clearly shows how efficiently these processing parameters affect the quality of Sandesh made from buffalo milk. Also chemical, microbial and rheological study of this new developed product was carried out. The specific finding and conclusions are as follows:

1. Buffalo milk which was standardized at 4% fat level resulted in more brittle, hard and coarse texture as compare to 5% milk. A minimum fat content of 5% in buffalo milk was essential to obtain Sandesh with satisfactory body and texture.
2. Addition of sodium citrate to different level (0.2% and 0.3%) prior to its heating and coagulation resulted in major enhancement in body and texture of buffalo milk Sandesh, though there was a problem of fat leakage. Out of the two different levels (0.2% and 0.3%) of sodium citrate used, product obtained at 0.3% was slightly more acceptable than at 0.2%.
3. Homogenization of buffalo milk in two stages at 1500 and 500 psi resulted in better body and texture without fat leakage and also it was close to control, whereas product obtained from buffalo milk homogenized at 1000 and 500 psi resulted in poor body, texture and overall acceptability.

4. For manufacture of Sandesh from buffalo milk, out of the two final heating temperatures, 90°c resulted better overall acceptability.
5. The yield of chhana was high (24.7%) from buffalo milk coagulated at 70°c followed by (21.9%) from buffalo milk coagulated at 80°c. The effect of temperature of coagulation was more conspicuous on body and texture of Sandesh. The body and texture of Sandesh samples made from buffalo milk coagulated at 70°c was better than coagulated at 80°c. As the coagulation temperature increase, the body becomes harder. There was little affect on the flavour and colour and appearance of the product due to temperature of coagulation.
6. When the mixture was cooked at 70°c for 20 minutes resulted in relatively soft Sandesh as compare to 80°c for 20 minutes which resulted in slightly hard, brittle and chewy product.
7. The sensory score for flavour, body and texture of final product was slightly more as compare to that of control whereas score for colour and appearance was comparatively less to that of control. There was no free fat on the surface of final product. The body and texture of final product was further improved than that of Sandesh made from previous individual experiments with homogenization and sodium citrate.
8. Sandesh made from control recorded more total solid (74.87%, 75.44% and 75.58%) as compare to that of buffalo milk Sandesh (73.58%, 74.14% and 73.77%). The fat and carbohydrate content decreased but the protein and ash contents increase slightly in the product from buffalo milk Sandesh. The titratable acidity of control was somewhat more to that of buffalo milk Sandesh.
9. The average total viable, yeast and mould and coliform counts/g of final products were17.7 $x10^3$, 5 x $10^1$and $2x10^1$ respectively.

10. There were significant reduction in the hardness, cohesiveness, springiness, gumminess and chewiness of the final product as compare to that of cow milk Sandesh.

Finally it was concluded that Sandesh made from buffalo milk using, 5% fat level, addition of 0.3% sodium citrate prior to heating and coagulation, homogenization of milk in two stages at 1500 and 500 psi, final heating temperature at 90°c, coagulation temperature at 70°c and cooking temperature70°c for 20 minute gave relatively better body and texture than that of cow milk Sandesh, whereas the colour and appearance was comparatively inferior.

BIBLIOGRAPHY

Ahmed, A. R., Vyas, S. H., Upadhyay, K. G. and Thakar, P. N. (1981). Study on manufacture of chhana from buffalo milk. *Gujarat Agri.Uni.Res.J.* **7**(1): 32-36.

Amerine, M. A., Pangborn, R. M. and Roessler, E. B. (1965). Principles of sensory evaluation of food, Academic press, Inc., New York.

Anantakrishna C. P. and Srinivasan, M. R. (1964). Milk products of India. ICAR, New Delhi cited by (Sahu, J. K. and Das, H. Chhana Manufacturing.) Mono graph of Indian dairy Association (2007). **3**:1-20.

Aneja, R. P. (1987). Keynote address of the national seminar on the "Recent advances in dairy processing", Karnal. April. p: 3

Aneja, R. P. (1997). Traditional dairy delicacies. A compendium in: P. R. Gupta (Ed), *Dairy India*, 5th Edn, New Delhi. Pp: 387-392.

Aneja, R. P., Mathur, B. N., Chandan, R. C. and Banerjee, A. K. (2002). Technology of indian milk products. *A Dairy India Publication*, New Delhi. Pp: 349-364.

Aneja, V. P., Rajoria, G. S. and Makkar, S. K. (1982). An improved process for continuous production of chhana. *Asian J. Dairy Res.* **1**(1): 41-44.

AOAC, (1995). Official methods of analysis, 16th Edn. Association of Official Analytical Chemists, Washington.

Bandyopadhyay P. and Khamrui K. (2007). Technological advancements on traditional Indian desiccated and heat acid coagulated dairy Products. *Bulletin of International Dairy Federation*. **415**: 4–10.

Bhattacharya, D. C. and Des Raj (1980). Study on the production of rosogolla- part-1: Traditional method, *Indian J. Dairy Sci.* **33**: 237-243.

Bourne M C (2002) *Texture profile analysis. Food Texture and Viscosity*, London: Academic Press. Pp: 182–188.

BSI (1969) IS-5162, Specification of chhana. BSI, 9, Bahadur Shah Zafar Marg, New Delhi.

Chakravarti, R.N. (1982). Dietary position of some Bengali Sweets. *J.Instn.Chem.* (India). **54**: 149.

Chaudhari, R. L., Berg, M. V. D., Singh, M. D. and Das, H. (1998). Effect of heat treatment on recovery of solid in chhana produced from cow and buffalo milk, *J. Fd. Sci. Technol.* **35**(1): 30-34.

Dairy India, (2007), Managing growth is the challenge. P. R. Gupta (Ed), six Edn, New Delhi. Pp: 15-42.

Das, H. (2000). Mechanized processing of Indian dairy product, chhana, paneer, sandesh and rasogolla. *Indian Dairyman.* **52**(12): 83-85.

Date, W. B., Lewis, Y. S., Johar, D. S. and Bhatia, D. S. (1958). Studies on the preservation and preparation of rasogolla. *J. Fd. Sci. Technol.* **7**: 217-220.

Davies, W. L. (1948). Indian indigenous milk product, Thacker Spink & Co. Kolkata. Pp: 61-71.

De, S. (1980). Outline of dairy technology. *Oxford university press*, New Delhi, Pp: 382-466.

De, S. and Ray, S. C. (1954). Study on indigenous methods of chhana making, *Indian J. Dairy Sci.* **7**: 113-125.

Gajendra, K. L. (1976). Effect of different concentration of citric acid coagulant on yield and quality of chhana from homogenized buffalo milk, M.Sc. Thesis submitted to Allahabad University, Allahabad.

Gira, V.K. (1978). Rheology of chhana from cow's and buffalo's milk, M.Sc. Dissertation submitted to Kurukshetra University, Kurukshetra.

Goal, V. K. (1970). Studies on manufacture and packaging of rosogolla. M.Sc. Dissertation submitted to Punjab University, Ludhiana.

Indian Standards, (IS: 1479, part II, 1961). Method of test for dairy industry. Chemical analysis of milk. Indian Standard Institution, Manak Bhavan, New Delhi.

IS (1969). Guidelines for cleaning and sterilizing dairy equipment IS 5253:1969. Bureau of Indian Standards, New Delhi.

IS (1977). Method of test for dairy industry. III. Bacteriological analysis of milk. IS: 1479, Part III, 1977. Indian Standard Institution, Manak Bhavan, New Delhi.

Iyer, M. (1978). Physico-chemical study on chhana from cow's and buffalo's milk. M.Sc. Dissertation submitted to Kurukshetra University, Kurukshetra.

Jagtap, G. E. and Sukla, P. C. (1973) A note on the factors affecting the yield and quality of chhana. *J. Fd. Sci. Technol.* **10**(2): 73-75.

Jagtiani, J. K., Iyenger, J. R. and Kapur, N. S. (1960). Studies on the preparation and preservation of rasogollas. *Food Science*, Mysore. **9**(2): 46-47.

Jana, A. H. and Upadhyay, K. G. (1990). The effects of homogenization conditions on the textural and baking characteristics of buffalo milk mozzarella cheese. *Dairy industry association of Australia*. **46**(1): 27-30.

Jankman, M. J. and Das, H. (1993). Optimization of process parameter for production of chhana from low fat cow milk. *J. Fd. Sci. Technol.* **30**(6): 417-421.

Kanawjia, S. K. (1975). Effect of homogenization pressures and fat level on the yield and quality of chhana from buffalo milk. M.Sc. Dissertation submitted to Kurukshetra University, Kurukshetra.

Kawal, S. (1979). Utilization of concentrated and dried milk for chhana making. M.Sc. Dissertation submitted to Kurukshetra University, Kurukshetra.

Khamrui, K. and Solanki, D. C. (2010). The relationship of textural characteristics with composition of sandesh produced from various market milk classes. *International J. Dairy Technol.* **63**(3): 451-456.

Kumar, G. and Shrinivasan, M. R. (1982). A comparative study on the chemical quality of three types of chhana samples. *Indian J. Animal Sci.* **52**: 741.

Kundu, S. S. and De, S. (1972). Chhana production from buffalo milk. *Indian J. Dairy Sci.* **25**: 159–163.

Ladkani, B. G. and Mulay, C. A. (1974). Feasibility of using Garber fat test for rapid estimation of fat in Khoa, *J. Fd. Sci. Technol.* **11**: 29.

Mahanta, K. C. (1964). Hand book of dairy science, Kitabistan, Allahabad. p: 386.

Makwana, A.K., Gurjar, M.D., Prajapati, J.P. and Shah, B.P. (2011). Opportunities for mini scale dairy enterprise. *Indian Dairyman*. **63**(1): 50-59.

Mini, G. S., Lily, G., Balasubramanian, S. C. and Basu, K. P. (1955). Quoted in (Rajorhia, G.S. and Sen, D.C. Technology of chhana: A review) *Indian J. Dairy Sci.* (1988). **41**: 141-148.

Patil, G. R. (2005). Innovative processes for indigenous dairy products. *Indian Dairyman.* 57: 82-87.

PFA, (1976). The Prevention of Food Adulteration Act. Eastern Book Company. Pp: 105.

PFA, (2006). The Prevention of Food Adulteration Act. 1954. Confederation of Indian Industries, New Delhi. p: 264.

Rahate, R. C. (1993). Effect of binding material from different sources on the quality of rosogolla. M.Sc. Thesis submitted to Dr. Panjabrao Desahmukh Krishi University, Akola (Maharashtra).

Rajorhia, G. S. and Sen, D. C (1987). Problem of milk sweets trade in India. *Indian Dairyman.* **39**: 283.

Rajorhia, G. S. and Sen, D. C. (1988). Technology of chhana: A review. *Indian J. Dairy Sci.* **41**(2): 141-148.

Rao, B. V. R. (1971). Effect of different concentration of citric acid coagulant and fat level of milk from crossbred cow on yield of chhana, fat losses in whey, moisture retained in chhana and quality of chhana. M.Sc. Thesis submitted to Allahabad University, Allahabad.

Ray, S. C. and De, S. (1953). Indigenous dairy products of India- III chhana. *Indian Dairyman.* **5**: 15.

Roy, S. K., Pandya, A. J., Patel, H. G. and Prajapati, J. P. (2010). Annual Report, SDAU, S.K.Nagar, Gujarat.

Sahu, J. K. (2007). Development of Process & Machinery for production of Sandesh on Indian dairy product. Unpublished PhD Thesis lIT. Kharagpur.

Sahu, J. K. and Das, H. (2007). Chhana Manufacturing. Monograph of *Indian dairy association.* **3**: 1-20.

Sahu, J. K. and Das, H. (2009). A continuous heat-acid coagulation unit for continuous production of chhana. *Assam Uni. J. Sci. Technol.* **4**(2): 40-45.

Sahu, J. K. and Jha, J. K. (2008). A storage stability of sandesh – an Indian Milk Sweet. Agricultural Engineering International: the CIGR *Ejournal*. 10:1-7. [Internet document]

Sanyal, M. K., Pal, S. C., Gangopadhyay, S. K., Dutta, S. K., Ganguli, D., Das, S. and Maiti, P. (2011). Influence of stabilizer on quality of sandesh from buffalo milk. *J. Fd. Sci. Technol.* **48**(6): 740-744.

Sarkar, J. K. (1975). Study on the composition of sandesh (Indian sweetmeat). *J. Fd .Sci. Technol.* **12**: 321.

Sen, D. C. (1985). Influence of calcium lactate strength on yield and sensory properties of chhana, *Asian. J. Dairy Res.* **4**(1): 36-38.

Sen, D. C. (1986). Effect of coagulation temperature on composition of calcium lactate chhana. *Indian J. Dairy Sci.* **39**(3): 244-246.

Sen, D. C. and De, S. (1984). Study on calcium lactate as chhana coagulant. *J. Food Sci. Technol.* **21**: 243.

Sen, D. C. and Rajorhia, G. S (1986a). Influence of pH of coagulation on chhana yield and quality using calcium lactate. *Asian J. Dairy Res.* **5**(1): 30-32.

Sen, D. C. and Rajorhia, G. S. (1985). Current status of sandesh production in *India. Indian Dairyman.* **37**: 4.

Sen, D. C. and Rajorhia, G. S. (1986b). Quality of Sandesh marketed in Delhi, *Asian J. Dairy Res.* **5**(4): 186-192.

Sen, D. C. and Rajorhia, G. S. (1990). Suitability of some packaging materials for packaging of sandesh. *J. Fd. Sci. Technol.* **27**(3): 156-161.

Sen, D. C. and Rajorhia, G. S. (1991). Production of narampak sandesh from buffalo milk. *J. Fd. Sci. Technol.* **28**(6): 359-364.

Sen, D. C. and Rajorhia, G. S. (1997) Enhancement of shelf – life of sandesh with sorbic acid. *Indian J. Dairy Sci.* 50(4): 261-267.

Sindhu, J. S., Arora, S. and Nayak, S. K. (2000). Physio chemical aspect of indigenous dairy products. *Indian Dairyman.* **52**(10): 51-64.

Singh M. D. (1994). Study on continues acid coagulation of buffalo milk. Unpublished PhD. thesis, Department of agriculture and food engineering, Indian institute of technology, Kharagpure, India.

Singh, G. P. and Ray, T. K. (1977). Effect of milk coagulants on the quality of rassogolla and sandesh. *J. Fd. Sci. Technol.* **14**: 149.

Singh, G. P. and Ray, T. K. (1977a). Effect of milk coagulants on the quality of chhana and chhana whey. *J. Fd. Sci. Technol.* **14**: 205-207.

Singh, R. S. and Mukhopadaya, S. (1975). Annual Report, NDRI, Karnal.

Snedecor, G. W. and Cochran, W. G. (1980). Statistical Methods: 7th Edn. The Iowa State University Press. Ames. IOWA, USA.

Soni, K., Bandopadhyay, A.K. and Ganguli, N.C. (1980). Manufacture of rassogolla from buffalo milk. *Indian J. Dairy Sci.* **33**: 357.

Tarafdar, H. N., Das, H. and Sitaram, P. (1988). Mechanical kneading of chhana & quality of rosogolla. *J. Fd. Sci. Technol.* **25**: 223-227.

Tilbury, R.H. (1980). Developments in food preservatives-I, Applied Sciences publishers Ltd., London. Pp: 1-25.

Verma, B. B. (1997). Technology of chhana based sweets in advances in traditional dairy products. CAS in Dairy Technology. NDRI Deemed University, Karnal. Pp: 64-70.

Wane, J. R. (1992). Role of yeast culture in the manufacture of rosogolla. M.Sc. Thesis submitted to Dr. Panjabroa Desahmukh Krishi University, Akola (Maharashtra).

CPSIA information can be obtained
at www.ICGtesting.com
Printed in the USA
LVHW101031300323
743042LV00003B/56

9 783659 585487